EFFECTIVE
TECHNOLOGY TRANSFER
IN BIOTECHNOLOGY
Best Practice Case Studies in Europe

EFFECTIVE **TECHNOLOGY** TRANSFER IN BIOTECHNOLOGY

Best Practice Case Studies in Europe

Editors

Oliver Uecke
Technische Universität Dresden, Germany
& Lipotype GmbH, Germany

Robin De Cock
Imperial College Business School, UK
& Ghent University, Belgium

Thomas Crispeels
Vrije Universiteit Brussel, Belgium

Bart Clarysse
Imperial College Business School, UK

Imperial College Press

Published by

Imperial College Press
57 Shelton Street
Covent Garden
London WC2H 9HE

Distributed by

World Scientific Publishing Co. Pte. Ltd.
5 Toh Tuck Link, Singapore 596224
USA office: 27 Warren Street, Suite 401-402, Hackensack, NJ 07601
UK office: 57 Shelton Street, Covent Garden, London WC2H 9HE

Library of Congress Cataloging-in-Publication Data
Uecke, Oliver.
　　Effective technology transfer in biotechnology : best practice case studies in Europe / Oliver Uecke, University of Technology Dresden, Germany & Lipotype GmbH, Germany, Robin De Cock, Imperial College Business School, UK, Thomas Crispeels, Vrije Universiteit Brussel, Belgium, Bart Clarysse, Imperial College Business School, UK.
　　　pages cm
　　Includes bibliographical references.
　　ISBN 978-1-78326-680-7 (hardcover : alk. paper)
　　1. Medical innovations--Europe. 2. Biotechnology--Technological innovations. 3. Technology transfer--Europe--Case Studies. I. De Cock, Robin. II. Crispeels, Thomas. III. Clarysse, Bart, 1969– IV. Title.
　　R855.5.E85U44 2015
　　610.28--dc23
　　　　　　　　　　　　　　　　　　　　　　　　　　　　　　　　2014047460

British Library Cataloguing-in-Publication Data
A catalogue record for this book is available from the British Library.

Copyright © 2014 by editors and authors
All rights reserved.

Typeset by Stallion Press
Email: enquiries@stallionpress.com

Printed in Singapore

This book presents best practice case studies on effective technology transfer in biotechnology in Europe (ETTBio). They were conducted during the ETTBio project.

The project ETTBio is co-financed by the European Regional Development Fund and made possible by the INTERREGIVC program.

Contents

List of Figures	xix
List of Tables	xxi
List of Contributors	xxiii
Introduction	xxvii
References	xxxiv

SECTION 1: TECHNOLOGY TRANSFER OFFICE (TTO)   1

Introduction	1
References	4

1. CASE STUDY 1: A Look Inside Imperial College's TTO   5

    1.1. Setting the scene   5
    1.2. Identifying best practices   7
        1.2.1. TTO team and structure   7
        1.2.2. Screening/assessment   11
        1.2.3. IP management   13
        1.2.4. Commercialization   14
    1.3. Implementing best practices in your region   16
    1.4. Future opportunities   17
    1.5. Best practices   18
    1.6. References   18

2. CASE STUDY 2: Technology Transfer at VIB   19

    2.1. Setting the scene   19
        2.1.1. VIB TTO   21

2.2. Identifying best practices 22
    2.2.1. The early beginnings and the catalyzing role of the Flemish government 22
    2.2.2. Screening and assessment process at VIB TTO 24
    2.2.3. IP management 27
    2.2.4. TTO support of entrepreneurs 28
    2.2.5. Results and achievements in terms of valorization 29
        2.2.5.1. Basic tech transfer statistics 29
2.3. Implementing best practices in your region 30
    2.3.1. Implications for TTO 30
    2.3.2. Implications for universities/research organizations 30
    2.3.3. Implications for policy makers 30
2.4. Future opportunities 30
2.5. Best practices 31
2.6. References 32

3. CASE STUDY 3: The Creation of a New Technology Transfer Office 33

    3.1. Setting the scene 33
        3.1.1. Introduction to IIMCB and Ochota Biocentre 33
        3.1.2. General information about the BioTech-IP 34
        3.1.3. History of technology transfer in Poland 34
    3.2. Identifying best practices 35
        3.2.1. BioTech-IP's history 35
        3.2.2. TTO team and structure 37
        3.2.3. TTO tasks and responsibilities 37
        3.2.4. Financing 38
        3.2.5. Achievements 38
        3.2.6. Future plans 39
    3.3. Implementing best practices in your region 39
        3.3.1. Implications for TTO 40
        3.3.2. Implications for research organizations 40
        3.3.3. Implications for policy makers 40

3.4.	Future opportunities	41
3.5.	Best practices	41
3.6.	References	42

4. CASE STUDY 4: A Model for IP Transfer and Shareholding for University Spin-Offs: The "Dresden Model" — 43

- 4.1. Setting the scene — 43
- 4.2. Identifying best practices — 46
  - 4.2.1. Evolution of the "Dresden Model" — 48
  - 4.2.2. The Dresden Model today — 48
- 4.3. Implementing best practices in your region — 49
- 4.4. Future opportunities — 49
- 4.5. Best practices — 49
- 4.6. References — 50

SECTION 2: FUNDING — 51

Introduction — 51
References — 54

5. CASE STUDY 5: Environmental Success Factors of Imperial College's TTO — 55

- 5.1. Setting the scene — 55
- 5.2. Identifying best practices — 57
  - 5.2.1. University level — 58
  - 5.2.2. Regional level: Financial community — 60
  - 5.2.3. Regional level: Government — 63
- 5.3. Implementing best practices in your region — 66
- 5.4. Future improvements — 68
- 5.5. Best practices — 68
- 5.6. References — 68

6. CASE STUDY 6: The Industrial Research Fund — 70

- 6.1. Setting the scene — 70
  - 6.1.1. Flanders — 71

6.1.2. The Flemish Department of Economy, Science and Innovation (EWI) 72
6.1.3. IWT 73
6.1.4. Vrije Universiteit Brussel 73
6.1.5. Interactions between universities and industry 73
6.2. Identifying best practices 74
6.3. The history and evolution of the IOF is closely linked to the Interface Offices at the Flemish universities 75
6.3.1. The workings of the Industrial Research Fund 76
6.3.2. Allocation key 77
6.3.3. Review and follow-up 78
    6.3.3.1. Strategic plan 78
    6.3.3.2. Information duty 79
    6.3.3.3. Monitoring and evaluation 79
    6.3.3.4. Review of progress 79
    6.3.3.5. The approach of VUB 79
    6.3.3.6. IOF Council 80
    6.3.3.7. VUB internal IOF regulation 81
    6.3.3.8. 2004–2011: Program funding 82
    6.3.3.9. 2011–…: Project funding 83
    6.3.3.10. Proof-of-concept funding 84
    6.3.3.11. Reporting 84
    6.3.3.12. Impact of IOF at VUB 84
6.4. Implementing best practices in your region 86
6.4.1. Implications for TTO 86
6.4.2. Implications for university/research organizations 86
6.4.3. Implications for policy makers 86
6.5. Future opportunities 86
6.6. Best practices 88
6.7. References 89

7. CASE STUDY 7: Regional Innovation Vouchers
   as an Effective Tool for Supporting Technology Transfer   90
   7.1. Setting the scene   90
   7.2. Identifying good practices   92
        7.2.1. Specific description   92
        7.2.2. Evolution of the innovation voucher   94
        7.2.3. The innovation voucher today   95
        7.2.4. The innovation voucher in the future   97
   7.3. Implementing best practice   100
   7.4. Implementing best practices in your region   101
        7.4.1. Implications for the implementation body
               (e.g. Agency, TTO)   101
        7.4.2. Implications for universities/research
               organizations   104
        7.4.3. Implications for policy makers   104
   7.5. Future opportunities   105
   7.6. Best practices   106
   7.7. References   106

8. CASE STUDY 8: Public Funds for Patenting,
   Valorization and Science–Industry Collaboration   108
   8.1. Setting the scene   108
   8.2. Identifying best practices   110
        8.2.1. Programs offered by National Centre
               for Research and Development   110
               8.2.1.1. Innovativeness creator   110
               8.2.1.2. Patent plus   111
               8.2.1.3. Applied research program   112
               8.2.1.4. SPIN-TECH Special
                        Purpose Vehicle (SPV)   112
               8.2.1.5. Operational Programme
                        Innovative Economy (OP IE)   113

8.3.	Implementing best practices in your region	115
	8.3.1. Application process	115
	8.3.2. Monitoring of projects	116
8.4.	Best practices	116
8.5.	Future opportunities	116
8.6.	References	117

SECTION 3: INCUBATORS — 118

Introduction — 118
References — 121

9. CASE STUDY 9: The Imperial Bioincubator — 123

9.1.	Setting the scene	123
9.2.	Identifying best practices	125
	9.2.1. Incubators at Imperial College	126
	9.2.2. Bioincubators in London	129
9.3.	Implementing best practices in your region	132
9.4.	Best practices	134
9.5.	Future improvements	134
9.6.	References	134

10. CASE STUDY 10: Idea Lab — A Platform for Students to Develop New Ideas — 136

10.1.	Setting the scene	136
10.2.	Identifying best practices	137
	10.2.1. Idea Lab	137
	10.2.2. Earlier initiatives and role models	138
	10.2.3. Foundation of the Idea Lab	139
	10.2.4. The working principles of the Idea Lab	140
	10.2.5. Outcomes	142
10.3.	Implementing best practices in your region	143
10.4.	Best practices	144

10.5.	Further opportunities	145
10.6.	References	145

SECTION 4: EDUCATION — 147

Introduction — 147
References — 150

11. CASE STUDY 11: Entrepreneurship and Technology Transfer Education at the Vrije Universiteit Brussel — 151

- 11.1. Setting the scene — 151
  - 11.1.1. Flanders — 151
  - 11.1.2. Vrije Universiteit Brussel — 152
  - 11.1.3. VUB TTO — 153
  - 11.1.4. Technology entrepreneurship at VUB — 156
- 11.2. Identifying best practices — 156
  - 11.2.1. Technology transfer interface — 156
    - 11.2.1.1. Booklet — 157
    - 11.2.1.2. Awareness creation, education and events — 157
    - 11.2.1.3. Technology days — 158
    - 11.2.1.4. Contract seminars — 159
    - 11.2.1.5. Posters, calendars and brochures — 159
  - 11.2.2. Technology entrepreneurship at VUB — 159
    - 11.2.2.1. Partners — 160
    - 11.2.2.2. Target group — 161
    - 11.2.2.3. Educational program — 162
- 11.3. Implementing best practices in your region — 164
  - 11.3.1. Implications for TTO — 164
  - 11.3.2. Implications for universities/research organizations — 164
  - 11.3.3. Implications for policy makers — 164
- 11.4. Future opportunities — 165
- 11.5. Best practices — 165
- 11.6. References — 166

12. CASE STUDY 12: BioEmprenedorXXI: Guidance Program for Starting up and Growing Companies in the Life Sciences Arena — 168

    12.1. Setting the scene — 168
    12.2. Identifying best practices — 170
        12.2.1. Organizational structure — 171
        12.2.2. Program contents — 171
        12.2.3. Phase I: Seizing opportunities — 171
        12.2.4. Phase II: Training entrepreneurs — 172
            12.2.4.1. Online training: Creating a business plan — 173
            12.2.4.2. Group face-to-face training: Creating a business plan and acquiring entrepreneurial skills — 173
            12.2.4.3. Individual face-to-face training (mentoring): Creating the business plan — 174
        12.2.5. Phase III: Creating a network of contacts — 174
            12.2.5.1. Networking lunches — 174
            12.2.5.2. Visits to technology parks and research centers — 175
        12.2.6. Phase IV: Preparing for investment and start-up — 175
            12.2.6.1. Ready for growth seminar — 175
            12.2.6.2. Healthcare Barcelona investment forum — 176
            12.2.6.3. BioEmprenedorXXI award: Presentation of business plans — 176
        12.2.7. Results of the program — 177
    12.3. Implementing best practices in your region — 178
    12.4. Future opportunities — 179
    12.5. Best practices — 179
    12.6. References — 180
    12.7. Acknowledgements — 180

13. CASE STUDY 13: Education for Scientists — 182
   13.1. Setting the scene — 182
   13.2. Identifying best practices — 184
      13.2.1. Educational courses for scientists provided by TTO — 185
         13.2.1.1. Course: "Workshop on scientific communication — interdisciplinary grants" — 185
         13.2.1.2. Course: "Everything you don't know about patents" — 185
         13.2.1.3. Course: "Research funding" — 186
         13.2.1.4. Course: "Effective communication and self-presentation" — 186
         13.2.1.5. Course: "Team management" — 186
         13.2.1.6. Course: "Negotiations with business partners" — 186
         13.2.1.7. Course: "Time management and personal effectiveness" — 187
         13.2.1.8. Course: "From the invention to the product — commercialization strategies in practice" — 187
      13.2.2. Science-to-business brunches — 187
      13.2.3. Scholarships for PhD students — 188
      13.2.4. Paid internships for PhD students and scientists in enterprises — 188
   13.3. Output/results/achievements — 189
   13.4. Implementing best practices in your region — 190
   13.5. Future opportunities — 191
   13.6. Best practices — 191
   13.7. References — 191

SECTION 5: CLUSTERS — 193
   Introduction — 193
   References — 195

14. CASE STUDY 14: The Biocat Model: Managing
the Bioregion of Catalonia 197

    14.1. Setting the scene 197
        14.1.1. Biocat 197
        14.1.2. The BioRegion of Catalonia 198
        14.1.3. Antecedents of Biocat 199
        14.1.4. Biocat's creation 201
        14.1.5. The first strategic plan 202
        14.1.6. The second strategic plan 203
    14.2. Identifying best practices 203
        14.2.1. Biocat's clear vision and mission 203
            14.2.1.1. Vision 203
            14.2.1.2. Mission 204
        14.2.2. Governance 204
        14.2.3. Core business 204
            14.2.3.1. Visibility/promotion 204
            14.2.3.2. Support in 205
        14.2.4. Main activities 205
            14.2.4.1. Cluster consolidation 206
            14.2.4.2. Business competitiveness
and talent 207
            14.2.4.3. Internationalization 209
            14.2.4.4. Social perception of
biotechnology 209
            14.2.4.5. KIC-IET 210
            14.2.4.6. B·DEBATE 210
    14.3. Implementing best practices in your region 210
    14.4. Future opportunities 211
    14.5. Best practices 212
    14.6. References 212

15. CASE STUDY 15: The Effects of a Cluster
on a Spin-Off — The Foundation of Ablynx 213

    15.1. Setting the scene 213
    15.2. Identifying best practices 215

15.2.1.	Trigger	215
15.2.2.	On patents and publications	216
15.2.3.	The top publication	218
15.2.4.	Government intervention	218
15.2.5.	The nanobody story continues	220
15.2.6.	Establishment of the company	221
15.2.7.	A story of people	222
15.3.	Implementing best practices in your region	223
15.3.1.	Implications for TTO	223
15.3.2.	Implications for universities/research organizations	223
15.3.3.	Implications for policy makers	223
15.3.4.	Implications for scientists	224
15.4.	Future opportunities	224
15.5.	Best practices	224
15.6.	References	225

16. CASE STUDY 16: Brokerage Event: Matching International R&D Projects — 227

16.1.	Setting the scene	227
16.2.	Identifying good practices	229
16.2.1.	Evolution of the Brokerage Event	230
16.2.1.1.	The Brokerage Event 2011	230
16.2.1.2.	The Brokerage Event 2012	231
16.2.1.3.	The Brokerage Event today	232
16.2.1.4.	The Brokerage Event in the future	232
16.3.	Implementing best practices in your region	233
16.3.1.	Implications for organizer (e.g. TTO)	233
16.3.1.1.	Event organization	234
16.3.1.2.	Other partners	234
16.3.1.3.	Involvement and contribution of the University Hospital Ostrava (strategic partner)	234
16.3.1.4.	Funding	234
16.3.4.	Implications for policy makers	236
16.4.	Future opportunities	237

16.5.	Best practices	238
16.6.	References	238

17. CASE STUDY 17: The DRESDEN-concept:
    A Focus on Shared Services and Facilities — 239

    17.1. Setting the scene — 239
    17.2. Identifying best practices — 240
        17.2.1. Specific description of shared facilities ("Technology Platform") — 242
        17.2.2. Evolution of the DRESDEN-concept for shared facilities — 243
        17.2.3. The "Dresden Model: Shared facilities" today — 243
        17.2.4. The DRESDEN-concept with shared facilities in future — 245
    17.3. Implementing best practices in your region — 246
    17.4. Future opportunities — 248
    17.5. Best practices — 248
    17.6. References — 248

Conclusion — 250
  Technology transfer offices — 250
  Funding — 251
  Incubators — 251
  Entrepreneurship education — 252
  Clusters — 253
  General conclusion — 253

List of Figures

Figure 1-1:	Overview of team experience and education.	9
Figure 7-1:	Innovation voucher diagram.	102
Figure 9-1:	Bioincubator in the Imperial College network.	128
Figure 9-2:	New research and translation campus in White City, West London.	129
Figure 9-3:	Overview of the most important bioincubators in the UK.	130
Figure 9-4:	London BioScience Innovation Centre.	130
Figure 9-5:	Queen Mary BioEnterprises (QMB) Innovation Centre.	132
Figure 12-1:	Phases of BioEmprenedorXXI program.	172
Figure 14-1:	Strategic roles of the Biocat organization.	205
Figure 14-2:	Biocat governance structure.	206
Figure 17-1:	Members of the DRESDEN-concept consortium.	241
Figure 17-2:	Services and facilities offered at the MPI-CBG.	244
Figure 17-3:	Screenshot of the web presence of the technology platform listing the DRESDEN-concept partners who already actively offer their equipment and the number of technologies they provide so far (as of 17/07/2013).	245

List of Tables

Table 2-1:	VIB statistics.	20
Table 2-2:	Basic TT statistics of VIB.	29
Table 7-1:	Timeline of the innovation voucher applications.	98
Table 11-1:	Course outline, "Business Aspects of Biotechnology", 2012–2013. Own set-up.	161
Table 11-2:	Overview of the technology entrepreneurship educational program at the Vrije Universiteit Brussel. Own set-up.	163
Table 12-1:	Characteristics of companies created.	178
Table 13-1:	Results of training activities.	190

List of Contributors

Robin De Cock

Department of Innovation and Entrepreneurship
Imperial College London Business School, UK

Department of Management, Innovation and Entrepreneurship
Ghent University, Belgium

Thomas Crispeels

Department of Business Technology and Operations (BUTO)
Vrije Universiteit Brussel, Belgium

Siim Espenberg

Department of Business Development
Tartu City Government, Estonia

Adela Farre

Biocat
Bariri Reixac, 4-8, Torre I (PCB)
08028 Barcelona, Spain

Alexander Funkner

Chair of Innovation and Entrepreneurship
Entrepreneurship network dresden|exists
Technische Universität Dresden
01062 Dresden, Germany

Marc Goldchstein

Department of Business Technology and Operations (BUTO)
Vrije Universiteit Brussel, Belgium

Tom Guldemont

Department of Business Technology and Operations (BUTO)
Vrije Universiteit Brussel, Belgium

Hubert Ludwiczak

Bio&Technology Innovations Platform
International Institute of Molecular and Cell Biology
Warsaw, Poland

Carlos Lurigados

Biocat
Bariri Reixac, 4-8, Torre I (PCB)
08028 Barcelona, Spain

Maarika Merirand

Department of Business Development
Tartu City Government, Estonia

Piotr Potepa

Bio&Technology Innovations Platform,
International Institute of Molecular and Cell Biology
Warsaw, Poland

Magda Powierża

Bio&Technology Innovations Platform
International Institute of Molecular and Cell Biology
Warsaw, Poland

Regional Development Agency Ostrava

Na Jízdárně 7
702 00 Ostrava, Czech Republic

Ilse Scheerlinck

Department of Business Technology and Operations (BUTO) –
Department of Business
Vrije Universiteit Brussel – Vesalius College, Belgium

Nadine Schmieder-Galfe

Chair of Innovation and Entrepreneurship
Entrepreneurship network dresden|exists
Technische Universität Dresden
01062 Dresden, Germany

Oliver Uecke

Chair of Innovation and Entrepreneurship
Entrepreneurship network dresden|exists
Technische Universität Dresden
01062 Dresden, Germany

Lipotype GmbH
Dresden, Germany

Introduction

This book presents 17 best practice case studies on "Effective Technology Transfer in Biotechnology". In recent years, a lot of effort and investment have gone into improving the outcome of technology transfer: the transfer of specialized patented or non-patented know-how from one organization to another (Zhao & Reisman, 2002). Technology suppliers are usually universities, research institutes or companies performing research and development (R&D). Customers demanding the technology are, for example, small and medium-sized enterprises (SMEs), large corporations and applied research institutes. In this book, we present an overview of initiatives that were deployed across Europe with the aim of supporting and stimulating the transfer of biotechnology discoveries and technologies. This book provides the reader with a critical assessment of the initiatives and with interesting lessons and inspiration for policy makers, entrepreneurs cluster managers and research institute managers when they are developing and implementing similar initiatives.

The origins of modern biotechnology go back to academic laboratories in the US and Europe in the sixties and early seventies. At these laboratories, scientists such as Stanley Cohen, Herbert Boyer, Georges Köhler and Cesar Milstein provided the technological and scientific breakthroughs that allowed for the development of novel biotechnology products. For instance, their work allowed for the development of the first *in vitro* recombinant therapeutic protein (insulin) by San Francisco-based Genentech. A major impetus for the development of the biotechnology industry was the passing of the

Bayh–Dole act in the US in 1980, which "allowed universities to own the patent arising from Federally funded research, and permitted them to grant exclusive licenses and to charge royalties, which would be shared with inventors" (Nelsen, 2005). Universities, now able to capitalize on these newly acquired assets, became the seedbed of many biotechnology start-ups and actively looked to license their technologies to industrial players. This process of valorizing academic research is now commonly known as technology transfer. The passing of the Bayh–Dole act ensured that the US biotechnology industry rapidly developed in the eighties and, as more products emerged from the biotechnology pipelines, that the industry took on additional functions besides R&D, such as production, market access and marketing (Fisken & Rutherford, 2002; Papadopoulos, 2000; Nosella et al., 2006).

Biotechnology is "the application of science and technology to living organisms as well as parts, products and models thereof, to alter living or non-living materials for the production of knowledge, goods and services" (Organization for Economic Co-operation and Development (OECD), 2005, p. 9).[1] It is referred to as one of the key enabling technologies of the 21st century and has the potential to offer solutions for health and resource-based problems the world is facing, such as unmet medical needs and fossil fuel dependency (European Commission, 2007a; OECD, 2009). As highlighted by Patzelt (2005, p. 1), the importance of biotechnology is "even more impressive if we consider that the modern biotech industry is only a quarter of a century old".

Applications of modern biotechnology are found in a range of different sectors (Link & Siegel, 2007). The pharmaceutical biotechnology sector is the most important biotechnology sub-sector, especially in terms of revenues and R&D spent (Luukkonen, 2005). The US has the biggest market share (65%), followed by the EU

---

[1] This broad definition is complemented with a list-based definition of biotechnology techniques which functions as an interpretative guideline to the single definition and focus on modern biotechnology applications. The biotechnology techniques used in this definition are: DNA/RNA; proteins and other molecules; cell and tissue culture and engineering; process biotechnology techniques; gene and RNA vectors; bioinformatics; and nanobiotechnology (OECD, 2005).

(30%) (European Commission, 2007a). Reports show that the US biotechnology industry is more developed than its European counterpart; for example in terms of number of biopharmaceuticals developed, number of companies, number of R&D employees and turnover (European Commission, 2007a; OECD, 2006). The pharmaceutical biotechnology industry has specific characteristics (Khilji et al., 2006; Hine & Kapeleris, 2006). Innovation is essential for success of biotechnology companies. These biotechnology innovations often possess a higher level of novelty compared to other industries. Due to this relatively high level of innovativeness, there exists a higher uncertainty with regard to the fact that these innovations will reach the markets. A major difference lies in the fact that academic research is a crucial source of innovation for the biotechnology industry. The increasing number of technology transfer activities within research institutes and universities exemplifies this (Branstetter & Ogura, 2005; Gross, 2009). Universities and research institutes have played a crucial role in developing technological innovations (Shane, 2004). Effective innovation and technology transfer processes in universities and research institutes are needed to develop innovations in biotechnology, which provide potential solutions to some of the global challenges faced by society (OECD, 2009a).

At the same time, challenges on a global scale are present. Fewer new drugs are getting market approval. Large multinational pharmaceutical companies have difficulties filling their R&D pipeline with new drug candidates, despite huge investments in R&D. Next to this, many drugs on the market will come off patent in the next few years, meaning that the future viability of pharmaceutical firms depends on the success of their R&D investments (Houlton, 2009; Frost & Sullivan, 2010b). In this light, a key challenge for companies is to identify and develop breakthrough innovations in-house, but also at an increasing rate with external partners from industry, government and academia (Frost & Sullivan, 2010a). However, results of technology transfer are not satisfying. Especially in Europe, this gap between academia and industry is often referred to as the "European Paradox" (European Commission, 2007b; Conti & Gaulé, 2011). In biotechnology, the translation of research results into new therapies is not satisfying due to existing barriers on academia, industry and government level (Albani & Prakken, 2009). To conclude, there exists a substantial interest for improving technology transfer at research institutes and specifically in the biotechnology context.

The EU project "Effective Technology Transfer in Biotechnology" (ETTBio) tackles these deficiencies existing within the knowledge and technology transfer processes. The project is co-funded by the European Union (ERDF – European Regional Development Fund) within the framework of the INTERREG IVC Programme. ETTBio is a three-year project with a total budget of 2.2 million euros and it addresses the challenging issue of transferring untapped economic potential of research institutes to the industry and to society at large. The project aims to identify, exchange and share best practices enabling successful and effective technology transfer in biotechnology, and to improve local and regional policies accordingly. In order to achieve this goal, ten project partners from seven EU Member States collected and analyzed best practices with regard to technology transfer in biotechnology. These partners include Technische Universität Dresden (Germany) as Lead Partner of the consortium, the City of Dresden (Germany), the Vrije Universiteit Brussel (Belgium), the Regional Development Agency of Ostrava (Czech Republic), the City of Tartu (Estonia), the City of Warsaw (Poland), the International Institute of Molecular and Cell Biology (Poland), Biocat (Spain), the Center for Genomic Regulation (Spain) and the Imperial College Business School of London (United Kingdom). This book presents best practice case studies on effective technology transfer in biotechnology in Europe. The selected case studies focus on the role of technology transfer offices (TTO), funding mechanisms, incubators, education and clusters. These cases were conducted during the ETTBio project.

All case studies presented in this book have been conducted using a multi-stage process. First, a conceptual framework was developed in order to evaluate the status of technology transfer in biotechnology as well as to identify strengths and weaknesses of each ETTBio partner region. Structured literature review on effectiveness of technology transfer as well as expert interviews were used to develop this framework. Opportunities for improvement of the technology transfer process were identified in the domains of TTO management, funding mechanisms, incubators, education and clusters. Then, best practice case studies in these domains were identified and drafted together with a group of experts.

The best practice case studies selected for this book start with a focus on the core organizations involved in technology transfer: TTOs, incubators and research institutes. With regard to the latter organizations, emphasis is put on case studies describing entrepreneurship and technology transfer education for students and researchers. We then look at funding mechanisms on organizational, regional, national and international level that facilitate the transfer of know-how and technologies from academia to industry. In a last step, we consider how technology transfer activities are embedded in a regional ecosystem. A thriving business or innovation ecosystem can contribute to success in technology transfer in biotechnology. The conceptual framework of these five fields is presented in Fig. 1.

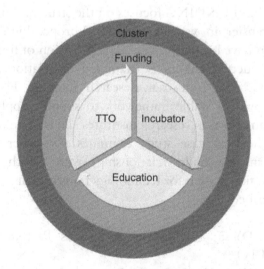

Figure 1: Conceptual framework for effective technology transfer in biotechnology.

The handbook is structured as follows:

In SECTION 1: TECHNOLOGY TRANSFER OFFICE, the role of TTOs as intermediate to manage technology transfer is discussed. Case studies on two well-established TTOs (Imperial College and VIB) reveal different ways in which these offices position themselves in relation to both government bodies and the industry. The examination of a newly formed TTO in Poland provides the reader with insights into the initial vision and strategy of a TTO and how a TTO makes its initial steps in the domain. The innovative "Dresden Model" is then presented as an alternative perspective on how technology transfer can be integrated into a university structure.

- CASE STUDY 1: A Look Inside Imperial College's TTO
- CASE STUDY 2: Technology Transfer at VIB
- CASE STUDY 3: The Creation of a New Technology Transfer Office
- CASE STUDY 4: A Model for IP Transfer and Shareholding for University Spin-Offs: The "Dresden Model"

SECTION 2: FUNDING focuses on the structure of funding for technology transfer in various parts of Europe, highlighting the impact this can have on the process. The provision of funding as one of a suite of success factors is examined in relation to Imperial College's TTO. The Industrial Research Fund in Flanders is an attempt to provide a stable framework to support applied research and valorization efforts at research institutes and is examined in CASE STUDY 6. The next case study examines a voucher system that encourages firms to build relationships with Czech universities. Finally, the funding system for technology transfer that is emerging in Poland is described.

- CASE STUDY 5: Environmental Success Factors of Imperial College's TTO
- CASE STUDY 6: The Industrial Research Fund

- CASE STUDY 7: The Regional innovation voucher as an Effective Tool for Supporting the Transfer of Technology
- CASE STUDY 8: Public Funds for Patenting, Valorization and Science–Industry Collaboration

SECTION 3 focuses on INCUBATORS. Incubators are structures that provide start-ups with office space, lab space and business services. They have proved to be highly successful structures in supporting biotechnology companies and in promoting technology transfer between universities and industry. The success of the existing incubator at Imperial College is extended by the plans to build a new incubator at Imperial's new site in West London. Guidelines for success in establishing an incubator are also given, as extrapolated from the experiences of Imperial College. Case Study 10 introduces the Idea Lab at the University of Tartu. The Idea Lab provides an incubation environment with mentors, resources and rewards for multidisciplinary student teams that want to try out new business ideas and want to build prototypes.

- CASE STUDY 9: The Imperial Bioincubator
- CASE STUDY 10: Idea Lab: A Platform of Idea Development for Students

The immense value of providing education for university researchers on the mechanisms of technology transfer is highlighted in SECTION 4: EDUCATION. The technology entrepreneurship project at VUB in Belgium is an initiative in which the university's business school and TTO work closely together to provide training to master and PhD students in both biotechnology and in business studies. A similar kind of "cross-fertilization" can be seen in the Spanish initiative BioEmprendedor XXI, in which businesses and research institutes work together to provide a bio-entrepreneurship training program that aims at starting up new biotechnology companies. Education in business for scientists is regarded as a key ingredient for technology transfer success in Poland as well, as assessed in CASE STUDY 13.

- CASE STUDY 11: Entrepreneurship and Technology Transfer Education
- CASE STUDY 12: BioEmprenedorXXI
- CASE STUDY 13: Education for Scientists

An effective way to foster technology transfer is to concentrate businesses and research institutions in a region so that they form a cluster. The ways in which such clusters have been established in Spain, Belgium and Germany are discussed in SECTION 5: CLUSTERS. This section also scrutinizes the structures each region has to support the needs and interests of technology transfer. In addition, the best practice case of "The Brokerage Event" in a Czech Cluster shows how universities, research institutions, and companies are successfully brought together in order to share new project ideas and in finding collaboration partners.

- CASE STUDY 14: The Biocat Model: Managing the Bioregion of Catalonia
- CASE STUDY 15: The Effect of a Cluster on Spin-Offs: The Ablynx Case
- CASE STUDY 16: The Brokerage Event: Matching International R&D Projects
- CASE STUDY 17: The DRESDEN-Concept: A Focus on Shared Services and Facilities

References

Albani, S. & Prakken, B. (2009). The advancement of translational medicine — from regional challenges to global solutions. *Nature Medicine,* 15(9), 1006–1009.

Branstetter, L. & Ogura, Y. (2005). Is academic science driving a surge in industrial innovation? Evidence from patent citations. *NBER WORKING PAPER SERIES,* W11561.

Conti, A. & Gaulé, P. (2011). Is the US outperforming Europe in university technology licensing? A new perspective on the European Paradox. *Research Policy,* 40(1), 123–135.

European Commission (2007a). Consequences, opportunities and challenges of modern biotechnology for Europe. *Analysis Report: Contributions of Modern Biotechnology to European Policy Objectives.* European Commission, Directorate-General, JRC Joint Research Centre, Institute for Prospective Technological Studies (Seville), Sustainability in Agriculture, Food and Health.
European Commission (2007b). *Improving Knowledge Transfer between Research Institutions and Industry across Europe.* Office for Official Publications of the European Communities, Luxemburg.
Fisken, J. & Rutherford J. (2002). Business models and investment trends in the biotechnology industry in Europe. *Journal of Commercial Biotechnology,* 8(3), 191–199.
Frost and Sullivan (2010a). *R&D, Innovation and Product Development Priorities: European Survey Results.* Growth Team Membership Research, Frost and Sullivan http://ww2.frost.com.
Frost and Sullivan (2010b). *Dynamics in the Pharma and Biotech Industry.* Pharmaceuticals & Biotechnology, Frost and Sullivan http://ww2.frost.com, 9837–9852.
Gross, C.M. (2009). Technology transfer: opportunities and outlook in a challenging economy. *The Journal of Technology Transfer,* 34(1), 118–120.
Hine, D. & Kapeleris, J. (2006). *Innovation and Entrepreneurship in Biotechnology, an International Perspective: Concepts, Theories and Cases.* Edward Elgar Publishing.
Houlton, S. (2009). Pharma refocuses on the patent cliff. *Chemistry World,* (6)1.
Khilji, S.E., Mroczkowski, T. & Bernstein, B. (2006). From invention to innovation: Toward developing an integrated innovation model for biotech firms. *Journal of Product Innovation Management,* 23(6), 528–540.
Link, A.N. & Siegel, D.S. (2007). *Innovation, Entrepreneurship, and Technological Change.* Oxford University Press, USA.
Luukkonen, T. (2005). Variability in organisational forms of biotechnology firms. *Research Policy,* 34(4), 555–570.
Nosella, A., Petroni, G. & Verbano C. (2006). Innovation development in biopharmaceutical start-up firms: An Italian case study. *Journal of Engineering and Technology Management,* 23(3), 202–220.
Nelsen, L.L. (2005). The role of research institutions in the formation of the biotech cluster in Massachusetts: The MIT experience. *Journal of Commercial Biotechnology,* 11(4), 330–336.

Organization for Economic Co-Operation and Development (2005). *A Framework for Biotechnology Statistics*. OECD Publishing, Paris, France.

Organization for Economic Co-Operation and Development (2006). *OECD Biotechnology Statistics 2006*. OECD Publishing, Paris, France.

Organization for Economic Co-Operation and Development (2009). *The Bioeconomy to 2030: Designing a Policy Agenda*. OECD Publishing, Paris, France.

Papadopoulos, S. (2000). Business models in biotech. *Nature Biotechnology*, 18(10), IT3–IT4.

Patzelt, H. (2005). *Bioentrepreneurship in Germany: Industry Development, M&As, Strategic Alliances, Crisis Management, and Venture Capital Financing*. Otto-Friedrich-Universität Bamberg.

Shane, S. (2004). *Academic Entrepreneurship: University Spinoffs and Wealth Creation*. Edward Elgar Publishing, Cheltenham, UK.

Zhao L. & Reisman A. (2002). Toward meta research on technology transfer. *IEEE Transactions on Engineering Management*, 39(1), 13–21.

SECTION 1: TECHNOLOGY TRANSFER OFFICE (TTO)

Introduction

A high number of discoveries and technologies applied in modern biotechnology are university-born. These discoveries and technologies serve as input resources for the biotechnology industry (Cockburn & Henderson, 1998; Whittington *et al.*, 2009). Recombinant DNA technology (UCSF and Stanford, US), the hybridoma technology (Cambridge University, UK) and transgenic plants (Ghent University, BE) are all examples of biotechnology discoveries and technologies originating from academic laboratories that were successfully commercialized. In the US, the Bayh — Dole act (1980) put universities in charge of pursuing and commercializing patents arising from government-funded research (Nelsen, 2005). This act effectively increased the rate of technology diffusion but meant that a new set of resources and capabilities had to be developed at academic institutions (Siegel *et al.*, 2003). The European counterparts of the Bayh — Dole act were put into place in the nineties. There exist a number of crucial differences between European and US universities, especially in terms of funding and patenting legislation. However, the increasing pressure to commercialize academic research led to increased patenting and commercialization activity at European research institutions (Markman *et al.*,

2008; Geuna & Rossi, 2011). Technology Transfer Offices (TTOs) are responsible for the assessment and protection of intellectual property as well as for technology transfer related issues of a university or research institute. A TTO is the interface between researchers and industry and can belong to the academic organization or can also be legally independent. The TTO can pursue a licensing-for-equity strategy, a sponsored research licensing strategy or a licensing-for-cash strategy, the latter being the most prevalent technology transfer strategy. Overall, royalty income and licensing appear to be overemphasized by TTOs (Markman et al., 2005).

Traditionally, we distinguish between two major routes that can be used to transfer biotechnology discoveries and technologies from the lab to the firm. Firstly, academic institutions can license their assets to existing organizations (licensing agreement). Secondly, academic institutions can license their assets to a new biotechnology firm (a spin-off) (Ahn & Meeks, 2008). This has led to the foundation of a large number of R&D-based small- and medium-sized enterprises (SMEs) that focus on the development of biotechnology applications and technologies coming from academic institutions. Consequently, there exists a need at the European research institutions to access knowledge on the commercialization of university-born inventions and to intensify industry contacts.

This means that the TTOs of academic institutions should be organized and staffed with professionals who have experience and knowledge in domains that do not form part of the academic core activities such as commercialization, production and development. Furthermore, a number of business skills such as prospection, project management and negotiation prove to be key in successfully transferring technology. To this end, universities are also increasingly seeking access to external complementary resources, such as knowledge on academic commercialization, by engaging in inter-organizational networks (Pusser et al., 2006). This networking function is often assigned to the institution's TTO.

Some institutions have reacted to this need for business skill by spinning off or outsourcing technology transfer activities. As such, we can distinguish between traditional university, non-profit TTOs and between for-profit TTOs. The latter structures are becoming more popular in Europe and appear to put an emphasis on spin-offs (Markman *et al.*, 2005).

The performance of a TTO is influenced by environmental and institutional factors and by organizational practiceks (Siegel *et al.*, 2003). Overall, TTOs exhibit low levels of absolute efficiency. TTOs are characterized by decreasing returns to scale and some authors have stressed the importance of the development of regional sector focused TTOs (Chapple *et al.*, 2005).

In this section, we introduce and describe three European technology transfer offices in Belgium, UK and Poland as well as an innovative model in Germany for creating university spin-offs in an environment where universities may not hold shares in spin-offs (Case Study 4). The three technology transfer offices present a nice mixture of experienced (Case Study 1 — Imperial College; Case Study 2 — VIB), and young (Case Study 3 — IIMCB) TTOs. This allows for the exploration of different compositions and strategies. We observe that a young TTO can — and probably should — act relatively pragmatically while strategies, procedures and infrastructure are already well-established in more experienced TTO structures. Furthermore, the selected case studies consist of two traditional non-profit TTOs (VIB and IICMB) and two for-profit TTOs (TU Dresden and Imperial). By contrasting these business models, we offer a full range of TTO options to interested policy makers and the governing bodies of academic institutions.

To summarize, the reader will observe how TTOs try to increase their efficacy and efficiency by specializing and by putting in place professional processes and structures. At the same time, TTOs are heavily influenced by their environment, especially by policy makers and of course by the scientific community they are servicing. It appears to be that government commitment and a long-term regional vision on R&D are necessary preconditions for TTOs to thrive and to develop.

References

Ahn, M.J. & Meeks, M. (2008). Building a conducive environment for life science-based entrepreneurship and industry clusters. *Journal of Commercial Biotechnology*, 14(1), 20–30.

Chapple, W., Lockett, A., Siegel, D. & Wright, M. (2005). Assessing the relative performance of UK university technology transfer offices: Parametric and non-parametric evidence. *Research Policy*, 34(3), 369–384.

Cockburn, I. & Henderson, R. (1998). Absorptive capacity, coauthoring behavior, and the organization of research in drug discovery. *The Journal of Industrial Economics*, 46(2), 157–182.

Geuna, A. & Rossi, F. (2011). Changes to university IPR regulations in Europe and the impact on academic patenting. *Research Policy*, 40(8), 1068–1076.

Markman, G.D., Phan, P.H., Balkin, D.B. & Gianiodis, P.T. (2005). Entrepreneurship and university-based technology transfer. *Journal of Business Venturing*, 20(2), 241–263.

Markman, G.D., Siegel, D.S. & Wright, M. (2008). Research and technology commercialization. *Journal of Management Studies*, 45(8), 1401–1423.

Nelsen, L.L. (2005). The role of research institutions in the formation of the biotech cluster in Massachusetts: The MIT experience. *Journal of Commercial Biotechnology*, 11(4), 330–336.

Pusser, B., Slaughter, S. & Thomas, S.L. (2006). Playing the board game: An empirical analysis of university trustee and corporate board interlocks. *The Journal of Higher Education*, 77(5), 747–775.

Siegel, D.S., Waldman, D. & Link, A. (2003). Assessing the impact of organizational practices on the relative productivity of university technology transfer offices: An exploratory study. *Research Policy*, 32(1), 27–48.

Whittington, K.B., Owen-Smith, J. & Powell, W.W. (2009). Networks, propinquity, and innovation in knowledge-intensive industries. *Administrative Science Quarterly*, 54(1), 90–122.

1

CASE STUDY 1: A Look Inside Imperial College's TTO

Robin De Cock
(Ghent University, Belgium;
Imperial College London, UK)

1.1. Setting the scene

Imperial College London is located in the heart of London and is a science-based institution with a reputation for excellence in teaching and research. Imperial College currently has the following three constituent faculties: Imperial College Faculty of Engineering, Imperial College Faculty of Medicine and Imperial College Faculty of Natural Sciences. The Imperial College Business School exists as an academic unit outside of the faculty structure. The Technology Transfer Office (TTO) is called Imperial Innovations. Imperial Innovations was founded in 1986 as the TTO for Imperial College London, to protect and exploit commercial opportunities arising from research undertaken at the College. In 1997, the Group became a wholly owned subsidiary of Imperial College London and in 2006 was registered on the Alternative Investment Market of the London Stock Exchange, becoming the first UK University commercialization company to do so.

Innovations has a Technology Pipeline Agreement with Imperial College London, which extends until 2020, under which it acts as the TTO for Imperial College London. The Group also acts as the TTO for Royal Brompton & Harefield, Chelsea and Westminster Foundation, Imperial College NHS and North West London Hospital Trusts. Following its £140 million fundraising in January 2011, the Group has invested larger amounts and maintained its involvement for longer in the most promising opportunities within its portfolio of spin-out companies, with the intention of maximizing exit values. In addition, the Group has made a number of investments in opportunities arising from intellectual property developed at or associated with Cambridge University, Oxford University and University College London, through its relationships with Cambridge Enterprise, Oxford Spin-out Equity Management and UCL Business. These universities are the top four research-intensive universities in Europe with a research income of over £1.3 billion per annum (Imperial Innovations, 2013).

In the beginning, Imperial Innovations was a Pro-Rector's Office and included 2–3 staff. From 1987 to 1997 the TTO grew very slowly and was mainly focused on IP management.

Like all early stage technology transfer offices, they were focused on filing patents and protecting stuff that looked very interesting. There was much more emphasis on that then there was on setting up companies. There were a few companies that were set up during that time but the college didn't make any money with that. (Penfold, 2012)

Between 1997 and 2003, the TTO team grew to a team of 20 persons and became a wholly-owned subsidiary of Imperial College London. Imperial College enhanced the incubation support and became more commercial. In 2004–2005, a private placement of shares resulted in £10 million for the College and £10 million for Imperial Innovations. In July 2006, the company completed its Initial Public Offering and was listed on the AIM of the London Stock Exchange, becoming Imperial Innovations Group plc. As a result of the IPO, the Group raised £26 million (and £30 million in November 2007) to invest in its portfolio of companies based on Imperial College London technology (Penfold, 2012).

Between 2008 and 2010, two of the Group's holdings were acquired in significant deals. In 2008, Pfizer acquired Group portfolio company, Thiakis, in a transaction valued at up to £100 million. The Group retained a royalty stream over future products arising from the Thiakis technology separate from the equity sale. In 2010, Janssen Biotech acquired Group portfolio company Respivert for an undisclosed amount. The Group received £9.7 million from the deal, a 4.7 times return on investment. In January 2011, the Group raised £140 million through a share placement. As part of this fundraising, it commenced investment in intellectual property developed at or associated with the Universities of Cambridge and Oxford and UCL, in addition to Imperial College London. The College still holds 30% of the shares.

During the last couple of years, Innovations invested significant amounts of money in their portfolio companies. In July 2011, the Group led a £40 million investment round for its portfolio company, Nexeon, a battery materials and licensing company developing silicon anodes for the next generation of lithium-ion batteries. One year later, the Group led a £72 million investment round for its portfolio company, Circassia, a specialty biopharmaceutical company focused on allergy. Also in 2012, the Group made significant investments in Cell Medica, an immune reconstitution specialist that raised £17 million. Recently, the Group was involved in a £20 million round for Mission Therapeutics and a £17 million round for Pulmocide (Imperial Innovations, 2013).

1.2. Identifying best practices

In what follows, we look at the TTO team and structure, the screening/assessment process, IP strategy and the commercialization strategy in order to identify best practices.

1.2.1. *TTO team and structure*

There is no "right" way to set up a TTO, but success does require considering some key issues. When we look at the TTO team and structure, Imperial Innovations makes a first distinction between TTO activities (technology transfer team) and investment activities (ventures team). Within the technology transfer team of Imperial Innovations, a second distinction is made between experts in the healthcare sector and experts in technology sectors.

The technology transfer team at Imperial Innovations is responsible for assessing, managing and beginning the commercialization process for intellectual property developed at Imperial College London, Royal Brompton & Harefield, Chelsea and Westminster Foundation, the Imperial College Healthcare NHS Trust and the Northwest London Hospitals Trust. The team manages the translation of an idea into practice, working with the inventor to secure translational or proof-of-concept funding and seeking input from potential industry partners to shape the opportunity. The technology transfer team will license opportunities to both larger corporations and small, growing companies. Together with the ventures team, it will also work on opportunities for the formation of new companies.

The ventures team at Imperial Innovations operates across the healthcare and technology sectors and is dedicated to building businesses. Where a company has been established, they will work to ensure the appointment of high-quality management, the development of strong relationships with industry, and will utilize their extensive operational expertise to provide the best possible platform for growth. With pre-company opportunities, they will thoroughly research and validate the opportunity, identify appropriate IP strategy, form the company, appoint motivated management and commit the necessary funding to build a future leading business. The ventures team invests selectively within the portfolio to accelerate commercial and technical development and to increase value, working alongside quality co-investors. They are an involved and active investor, preferring to be involved at the earliest possible opportunity. In addition, the ventures team has made a number of investments in opportunities arising also from intellectual property developed at or associated with Cambridge University, Oxford University and University College London, through its relationships with Cambridge Enterprise, Oxford Spin-out Equity Management and UCL Business. The venture team is supported by a group of venture partners.

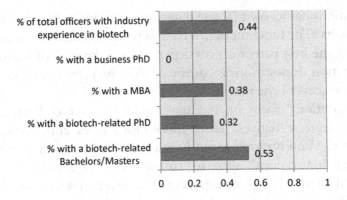

Figure 1-1: Overview of team experience and education.

Imperial Innovations has a workforce of 34 TTOs (technology transfer team) and investment managers (ventures team). Figure 1-1 gives an idea of the experience, education and background of these persons. We can conclude that industry experience, a PhD in sciences and an MBA are important if you want to become a part of the Innovations team (Imperial Innovations, 2013).

The healthcare group consists of 18 FTEs: 61% has a PhD in biotech related fields; 44% has an MBA; and 83% has industry experience in biotech related fields. The biotech related industry experience in the healthcare group is around 11 years, while the average experience as TTO is 5 years. In the ventures team, we noticed the extensive experience in big venture capital funds like 3i. Overall, industry experience seems to be a very important quality required of Innovations team members. In order to attract these industry experts, Imperial Innovations pays their officers more than the usual university wages. To further convince industry experts to join Innovations, they also offer them non-material rewards by giving them responsibilities, a real challenge and respect. These industry experts can earn much more money in their sector so they need to be convinced with a compelling offer, which includes both financial and non-financial rewards.

These two teams are supported by entrepreneurs in residence, a network of entrepreneurial CEOs and an IP due diligence team.

Innovations has its own IP due diligence team. However, they do not file patents. The function of this team is to screen patent agencies and pick out the best patent offices depending on the type of technology or invention. Innovations is convinced that the patent filing is crucial for the success of the patent.

Innovations' focus on industry experience is also illustrated by two concrete initiatives. First, Innovations hires entrepreneurs in residence. The Entrepreneur in Residence (EiR) program is a part of Imperial Innovations' aims to ensure that world-class science is matched by world-class management. The program is well established. In practice, an EiR will spend a period, typically two days a week over the space of three months, located at Imperial Innovations' offices. They will appraise existing technologies in the portfolio and meet with Imperial academics with the intention of selecting a technology or family of technologies that they feel can address a lucrative market need and around which a suitable IP protection strategy can be developed. They will then work with the ventures team at Imperial Innovations to build a business proposition and obtain start-up funding, with Imperial Innovations typically co-investing. They will attract a skilled and experienced management team to advance the technology through the stages needed to successfully bring a product to market. Once seed-funding has been secured the company may move to the Incubator, where it can develop into a mature business. Imperial Innovations will continue to support the business through necessary funding rounds.

Second, Innovations also has a network of entrepreneurial CEOs. Entrepreneurial CEOs are experienced, first-rate executives with leadership qualities and excellent experience either in building and selling technology companies or in the creation of significant value for large corporations, who will partner with Innovations to identify and build the next generation of companies. They combine technical and commercial competence and bring ideas of where they would like to form a new company. Imperial Innovations is always looking to identify people looking to build their next business in the areas of healthcare, energy or engineering.

Finally, the ventures team is supported by a group of venture partners. These are experienced industry experts or entrepreneurs who execute due diligence, scope out investment opportunities, help in recruiting the right people and offer a network.

1.2.2. Screening/assessment

The TTO team faces different challenges if they want to make their technology transfer process work. Traditional technology transfer processes start with screening and assessment. However, a very important step, that is often forgotten, is the development of a trusting relationship with the academic community to bridge the cultural divide between university and industry and complement the knowledge and network of the academic community.

> *Academics at the college, they do not think of us in first place, they think of getting publications and getting the next grant.* (Director Entrepreneurship Hub, Imperial College Business School, November 2013)

The TTO works hard to maintain good relations with the inventors and academics at the college. Executives from Innovations have regular information meetings with research groups. The TTO is involved in lots of courses, seminars and other sessions for undergraduates, graduates, PhD students and post-doctorates. There are many one-to-one meetings between academics and Innovations staff. This is again crucial to avoid people publishing before they want to file a patent. Most of the time, researchers propose their inventions to Imperial Innovations. Although the ventures team sometimes identifies interesting investment opportunities in particular areas and asks the technology transfer team to go and talk with the departments at Imperial College in that area, the relatively small TTO team — compared to the research community of Imperial College — is often not able to proactively screen every department at Imperial College on interesting technologies. That's why it is important to build up a strong image and be involved in different meetings, courses and events at the College.

Initial technology screening is largely carried out in-house, but external experts may be brought in for some technologies. The Entrepreneurs in Residence and the network of entrepreneurial CEOs are sometimes consulted.

> *These entrepreneurs work quite focused, they have their specific expertise while the TTOs respond to every request of every researcher at the college.* (Deputy Director, Imperial Innovations, November 2012)

Seeking feedback from potential customers in this initial phase is often too early, especially in the biotech context. It could also lead to premature disclosure of inventions and thus damage patent prospects. Innovations does its own market research and also commissions external market research, where required. External venture capitalists are not yet interested in this stage of development because there is often no concrete return and the technology is still far away from the market.

The screening and assessment of invention disclosures gets a lot of attention. The selection criteria are clear and straightforward and are based on two simple but vital questions: is it protectable and is it worth protecting? Attempting to quantify precisely the future value of such early-stage technology is rarely carried out. In the biotech space, the TTO's staff frequently have a good idea of whether the invention meets its value-threshold without detailed research. In some specialized cases, more detailed market research will be carried out, either in-house or externally. Imperial Innovations also gives inventors/researchers who are not selected (around one out of eight are selected) extensive feedback and explain to him/her why there are not interested in filing a patent for his/her invention.

> *Assessing an invention is something we already do instinctively when we talk to researchers: Where does this fit in the market and in the industry? What is already out there? And how are we going to commercialize it? ... The next steps really depend on what kind of inventions you are talking about. In some cases, you do not have to do much more market research to*

know that these inventions will not work. In other cases, if it is a new drug for instance ... we quickly know/identify who the leading companies are in that domain and we will contact them. If you are talking about a certain component which fits in a certain engineering process then it becomes more difficult to assess the potential. In this case, we will probably do some more market research to investigate whether or not there is a clear need for this invention. (Deputy Director, Imperial Innovations, November 2012)

The decision whether to seek to out-license the technology or to incorporate into a spin-out is frequently a complex one. Some of the many criteria to be considered are: 1. What is the scalability/size of the market? 2. Is the technology stand-alone or does it need complementary assets and other technologies? 3. What are the possible exit strategies? 4. What does the inventor want?

1.2.3. IP management

Besides screening and assessing the technology, it also important that the TTO is able to manage the developed IP. Patentability and likely Freedom to Operate are always evaluated before patenting. Potential licensees are checked whenever the Imperial Innovations is able to do so. They also try to plan the commercialization strategy and technology roadmap before filing. A clear patent strategy seems to be crucial but, under time constraints and pressures, industry experience is equally important. This is illustrated by the quote from a technology transfer officer in the healthcare technology transfer team, Imperial Innovations:

Whenever we can we try to plan the commercialization strategy and technology roadmap. We also screen potential licensees. However in some cases where someone is about to make something public, it is not possible. In these cases, we will have to act quickly and rely on our industry experience. (Deputy Director, Imperial Innovations, June 2013)

There is a clear patent strategy which covers "to file or not to file", timing, geographical scope, patent enforceability, cost containment and the communication with national, European and US patent offices. The inventors are closely involved in this process. At each stage, the responsible technology transfer officer will have to justify the costs and get approval from the budget holder(s). Innovations pays all the patent costs and only recovers those costs from the inventions successfully commercialized.

Imperial innovations also has an IP due diligence team that does not file patents but searches for the patent offices which are most appropriate for filing the specific invention or technology. The head of the team is a fully qualified UK and European Patent and Design attorney with 12 years working experience as an in-house patent attorney.

1.2.4. *Commercialization*

The formation of the commercialization strategy often happens in collaboration with the inventor. Innovations draws upon its own staff's expertise and networks to provide important feedback in the development of the commercialization strategy. If the technology is deemed to be at too early a stage, yet it has potential, the TTO may assist the investigator in finding funding or partners for further development.

The ways of commercializing depend largely on whether the field is therapeutics, diagnostics, biofuel, biochemical inventions or bioengineering. When it concerns biotech therapeutics or diagnostics, Innovations has a clear network of potential partners. Based on the invention, they generally have a good idea of which companies need to be contacted and they also have the contact details of the right person in these companies. If that doesn't work, Imperial Innovations tries to contact smaller biotech companies. A TTO from the technology transfer team explains:

> We contact them with a datasheet and say this is what we've got, it's at this stage and these are the patent details. Crucially, we seek to show the value of the technology or product to the company's business. In this case, Innovations is the principal contact with the company. Occasionally the

inventor already has established links with a company — in which case we'll build on those links. (Deputy Director, Imperial Innovations, June 2013)

The recently renewed website of Innovations also includes summaries of inventions and technologies in order to attract potential licensees and commercialize the inventions (Imperial Innovations, 2014).

In this stage of the commercialization process, the venture team of Innovations plays an active role and is involved in every important meeting. A member of the venture team explains:

> *The venture team works closely together with the tech transfer team and will support them in this stage by either looking for companies who could potentially be interested in using/accessing the technology and setting up the license or royalty stream or — if there is even more potential — they will help to set up a spin-off venture by composing a strong entrepreneurial team, write the business plan, connect them with industry experts and find/provide funding. (Bhaman, 2012)*

> *The venture team has a lot of investment experience and are assisted by a group of venture partners. These are experienced industry experts or entrepreneurs who execute due diligence, scope out investment opportunities, help in recruiting the right people, offer a network. For instance, X is former top manager of the company VISA, he is very useful in transferring everything that has to do with payment software/technologies. (Deputy Director, Imperial Innovations, June 2013)*

Imperial Innovations also offers incubation space for biotech companies in the Imperial College bioincubator (see Case Study 9) and manages his own fund. The ventures team of Innovations identifies and invests in business opportunities arising from Imperial College and also from Cambridge, Oxford and UCL. Investments are made from the balance sheet of Innovations. There is no separate "fund" in the sense that many venture capital funds operate. This gives Innovations the freedom to invest according to its own criteria and means it is less subject to the whims, prejudices and fashions of the "market".

> *Innovations is not constrained by the VC mentality where they have to raise and spend money at time x and realize a return by time y. Innovations can take the long view if they need to. We can invest for 2 years in a company but we can also choose to invest for 10 years.* (Director of Healthcare Investments, Imperial Innovations, May 2013)

Innovations' equity portfolio consists of 93 companies (31.01.2014). A negative funding decision by Innovations does not automatically jeopardize a spin-out's chances of raising money from other investors. A quote from different members of the Imperial Innovations investment team illustrates this:

> *Getting a "No" from Imperial Innovations does not make it impossible for spin-outs to raise money from external investors. It is not the case that because Imperial Innovations does not like it, other investors will automatically refuse investing in this invention.* (Director of Healthcare Investments, Imperial Innovations, May 2013)

1.3. Implementing best practices in your region

In this case study, we focus on best practices in the internal processes of the TTO. During the interviews with TTOs, a few guidelines were suggested in order to develop a strong technology transfer system. When you only look at the level of the TTO then human and financial capital are two crucial success factors for a TTO. We have already shown how Imperial Innovations is structured and how important it is to attract people with industry experience and/or entrepreneurial experience is crucial. However, it is not easy to attract these profiles. Imperial Innovations managed to attract this human capital through different actions. First, they tried to offer a competitive package of financial incentives. Secondly they also offered various non-financial incentives. Some industry experts were attracted by Imperial's offer because it included a new intellectual challenge with lots of variety and a fast moving environment. It is important that the TTO gives them a challenge, responsibility and respect. Third, if it is difficult to hire them in the TTO team, one can choose to find different ways to benefit from industry experts and experienced entrepreneurs. Examples in the case of Imperial Innovations are the entrepreneurs in residence program and the pool/network of entrepreneurial CEOs.

The best practices mentioned in this case study need to be implemented with great care and with a good knowledge of the universities' structure and ecosystem. In other words, they need to be aligned with the universities' structure and match the ecosystem of the university and the region. Only then they could also work in different regions and universities.

> *Everyone who wants to replicate us, has to do it differently simply because the institution is different and the investment environment is different.* (Director of Healthcare Investments, Imperial Innovations, May 2013)

Finally, the best practices identified in the technology process of Imperial College will only work if there is sufficient qualitative input and the "raw materials" delivered by a strong university. As one the TTOs stated:

> *You can't make a silk purse out of a sow's ear. We do the best we can with the raw material we get. If nobody invents a blockbuster at Imperial, Innovations does not have a blockbuster to commercialize. The technology transfer team depends on the inventions of Imperial College researchers while the ventures team depends on the investment opportunities at Imperial, Cambridge, Oxford and UCL.* (Deputy Director, Imperial Innovations, June 2013)

1.4. Future opportunities

Imperial College structures are mainly applicable for technology push ideas coming from the academic community at the College. Therefore, it started to build an accelerator and central hub to support more market pull ideas that do not necessarily come from the academics only.

1.5. Best practices

- Human capital TTO: industry and entrepreneurial experience (pool of entrepreneurs, entrepreneurs in residence, venture partners ...)
- Internal structure TTO: technology transfer team vs ventures team
- Development of a trusting relationship with the academic community
- Internal patent attorney
- Own fund (investing from their own balance sheet)

1.6. References

Papers, reports, books, conference presentations and websites

Bhaman, M. (2012). *How Innovations Selects, Sets up and Manages its Spinout Companies.* ETTbio site visit, Imperial College London.

Imperial Innovations (2013). Retrieved from: http://www.imperialinnovations.co.uk. Accessed on 12.01.2013.

Imperial Innovations (2014). Retrieved from: http://www.imperialinnovations.co.uk. Accessed on 22.01.2014.

Penfold, H. (2012). *How Innovations Started and Grew to What it is Now.* ETTbio site visit, Imperial College London.

Case study interviews

- 16.11.2012, Deputy Director, Technology transfer team, Imperial Innovations.
- 28.03.2013, Member of venture team, Imperial Innovations.
- 13.05.2013, Director of Healthcare Investments, Imperial Innovations.
- 21.06.2013, Deputy Director, Technology transfer team, Imperial Innovations.
- 20.11.2013, Director of Entrepreneurship Hub, Imperial College Business School.

2

CASE STUDY 2: Technology Transfer at VIB

Tom Guldemont, Thomas Crispeels and Ilse Scheerlinck
(Vrije Universiteit Brussel, Vesalius College, Belgium)

2.1. Setting the scene

The Flanders Institute for Biotechnology (VIB) is a life sciences research institute in Flanders, Belgium. With around 950 scientists from over 60 countries, it performs basic research into the molecular foundations of life. VIB is an excellence-based entrepreneurial institute that focuses on translating basic scientific results into pharmaceutical, agricultural and industrial applications.

VIB works in close partnership with four universities — UGent, K.U.Leuven, University of Antwerp and Vrije Universiteit Brussel and is funded by the Flemish government. On these four campuses, and therefore geographically spread over Flanders, VIB unites eight departments, 70 research groups and six service facilities.

VIB, which arose out of the government's desire to support the life sciences sector in the Flemish region and was set up in 1996, receives extensive government funding: yearly 38.1 million euros for the period 2007–2011, marking a 20% increase over the preceding five-year period. The Flemish government's investment in VIB is contractually stipulated in five-years-management agreements, so-called "convenants". This large Flemish government investment has some strings attached though. As stipulated in the management agreement, the Flemish government has very precisely described expectations of this investment, in terms of scientific productivity and industrial and social valorization. That is why the Flemish government thoroughly evaluates VIB every five years. It examines whether the preset objectives have been realized and whether the benefits for Flanders justify such an investment.

VIB has around 1,300 faculty members including 950 research employees, of which 414 PhD students in 2011. As VIB is focused on biotechnology, all employees are active in this field. It is important to mention however that a small majority of these employees are still on the payroll of universities.

VIB is a pure research institute. It provides no education at all and has therefore no (under)graduate student subscriptions. Neither has the institute a business school.

Table 2-1 presents some basic employment and revenue statistics of VIB.

Table 2-1: VIB statistics. *Source*: D'Hondt, 2012.

	2006	2007	2008	2009	2010
Employment by VIB (31/12)	435	454	471	515	504
Employment by universities (31/12)	561	613	624	710	737
Employment total	996	1.067	1.095	1.225	1.241
Turnover (k€)	8.805	11.240	20.660	12.200	14.455
Subcontracting (k€)	1.650	723	-330	4.116	3.021
Grants and subsidies (k€)	36.436	40.712	40.622	45.786	50.495
Other revenue (k€)	442	270	676	432	542
Company revenue (k€)	47.337	52.945	61.628	62.534	68.513
Value purchased products and services	5.598	6.095	6.086	7.058	7.190
Added value (k€)	5.303	6.138	14.920	9.690	10.828
Company revenue per VIB FTE (k€)	108,8	116,6	130,8	121,4	136,0
Added value per VIB FTE (k€)	12,2	13,5	31,7	18,8	21,5

2.1.1. VIB TTO

The Technology Transfer Office (TTO) of the VIB is a centralized office, located in the headquarters of the institute. The VIB TTO employs 16 people, of which 12 technology transfer officers and four administrative staff members. The 12 officers are composed of four IP managers, two of whom are patent attorneys, four licensing managers, two business developers, one contract manager and one general manager, who is also involved in business management.

The mission of the TTO is to translate research into products, to promote economic growth, to create jobs and to generate research funding, in that order.

As already mentioned, the institute has eight biotech faculties or research departments that are being served by the TTO. Additionally, the TTO is also offering its services to biotech research groups that are not part of VIB. This can be ad hoc, but there is also structural support to the University of Hasselt, the University of Antwerp and the University of Ghent. VIB TTO is also providing technology transfer (TT) services to foreign TT offices, such as Institute Spain and CRP Santé in Luxembourg.

For the TTO, it is highly important that the VIB research results are translated for the benefit of mankind and that they reach the market (socioeconomic impact 80%). Of course, excellent science is the base of everything and you need to keep the researchers motivated, but academic satisfaction is only of minor importance for the TTO (20%). The TTO is not profit-driven, therefore it attaches no importance whatsoever to profit, which means that all revenues are reinvested in the institute (Leyman, 2012).

The VIB is evaluated according to several performance criteria, which are imposed in the covenant by the Flemish government to justify its investment in the institute. Among these performance criteria are generated revenues, the number of patents and the number of start-ups that are dependent on the licensing of VIB technology. Parameters that are not part of the performance criteria are for example profits (logically, as the VIB is a non-for-profit organization), the number of actual licenses and the number of products on the market.

As mentioned, the TTO is comprised of 12 full-time officers with a diversified background. Eight of them already gained experience in the biotech industry, while one person has a legal education. None of them is a Certified Licensing Professional, has a business PhD or an MBA. On the other hand, all of them have a biotech-related master's degrees and pursued a biotech-related PhD.

The officers that already have biotech industry experience, spent on average two years in those companies. In terms of tech transfer activity, the officers have on average six years of experience. The impression that this number is not that high can be explained by the fact the VIB and its TTO are growing very rapidly and that by consequence new members with little experience have joined the group in recent years. Nevertheless, it has to be mentioned that six years of experience is in fact not low, given the fact that all these people have a biotech PhD and some have additional degrees. As a consequence, new members are not really 'junior' starters (Leyman, 2012).

VIB has two patent attorneys in-house and is very satisfied with the work they are doing.

2.2. Identifying best practices

2.2.1. *The early beginnings and the catalyzing role of the Flemish government*

The biotechnological excellence and successes of the early eighties in Flanders attracted the attention of the regional policy makers. In 1989, the Flemish government launched the VLAB-program (Vlaams Actieplan Biotechnologie, Flemish Action Plan Biotechnology) which was specifically designed to detect and support promising biotechnology projects. Remarkably, the focus of the agency was towards valorization, a revolutionary idea at that time. Rudy Dekeyser and his colleague Jo Bury were both scientific consultants at the Agency for Innovation through Science and Technology (IWT, set up in 1991) and responsible for the VLAB program. In 1993, they were both invited to the office of Luc Van den Brande, then Minister-President of the Flemish government. Wanting to put the Flemish biotechnology industry on the world map, Van den

Brande asked them to provide recommendations as well as a plan to accomplish it. His exact words were:

> "I want to invest 1 billion Belgian francs [25 million euros] annually to put the Flemish biotechnology on the world map. Excellence in research has to be the focus, but the translation to social and economic value is at least as important. Here is a white sheet of paper, to you the task to forge a plan." (VIB, 2012a; VIB, 2012b)

Dekeyser and Bury immediately had a new "institution" in mind; they didn't merely want to create a new "granting body" because this approach held the risk of spreading the government funding too thin. Together with government employee Dirk Callaerts, Bart De Moor and Christine Claus, Director General of IWT, Jo Bury and Rudy Dekeyser formed a team that gradually elaborated the plan. In February 1994, less than ten weeks after their first meeting, Dekeyser and Bury returned to the Minister-President's office. The first plan was unfolded, and Van den Brande saw that it was good. He sent Dekeyser and Bury — discretely — to the field to check the reaction of stakeholders to the plan.

Some were very enthusiastic while others responded much more cautiously. In April 1994, the plan was officially presented and discussed in the parliament and the Flemish government. The period between April 1994 and April 1995 was quite bumpy with much debate on the topic. One of the main concerns of the academic community was that only two of the four Flemish universities would be present in this new biotechnology structure/institute. Meanwhile, the Minister-President was becoming impatient. A general election was due to be held 21 May 1995 and he wanted to establish his biotechnology brainchild before this happened.

On April 5 1995, during its last meeting, the Flemish government approved the establishment of the VIB. The new initiative was a research institute with four core departments, more specifically, the labs from Herman Vanden Berghe and Desire Collen (both KULeuven), and Walter Fiers and Marc Van Montagu (both Ghent University). These core departments were supplemented with five associated departments: the labs of Nicolas Glansdorff

(Vrije Universiteit Brussel), Raymond Hamers (Vrije Universiteit Brussel), Danny Huylebroeck (KULeuven), Christine Van Broeckhoven (University of Antwerp) and Joël Vandekerckhove (Ghent University). In fact, the nanobody know-how (see Case Study 15) seemed to be crucial in order to incorporate a VUB research department in the VIB.

VIB became an institute without walls, in which the research groups remained on their university campuses. To enable this, framework agreements were signed between all universities involved and VIB. VIB also set up an administrative headquarters in Ghent. To achieve the objectives, the Flemish government provided an annual grant of 920 million Belgian francs (23 million euros) to the institute.

At inception on January 1, 1996, Jo Bury became Managing Director of VIB, while Rudy Dekeyser became Deputy Director and Director Technology Transfer (VIB, 2012a; VIB, 2012b).

2.2.2. Screening and assessment process at VIB TTO

The VIB TTO has a relatively large pool of inventions from which to identify and select an average of 50 inventions per year. In 2010, VIB scientists generated 57 inventions. This brings the total of reported inventions since VIB's inception until 2010 to 689. Approximately 50% of these inventions are protected by a patent application.

To screen the technologies or inventions in the VIB, the TTO does not rely on external groups. In other words, all screening is performed internally. To assess the identified inventions the TTO sometimes asks for feedback from specialized life sciences consultants and potential customers of the technology, but it does not ask the opinion of industry veterans or venture capitalists. Although it is rather difficult to estimate how much time is spent by the TTO on screening and assessment, 30% of the time for all activities seems a reasonable answer. This is calculated by considering the number of people that are focused on screening and assessment and comparing that number with the total number of TTO officers.

The relationship between the TTO and the inventor or research group that is responsible for the invention is rather straightforward.

The TTO clearly is in the "driver's seat" and decides how the valorization path is continued (as it is requested in the covenant with the Flemish government). There are in fact no differing goals. If research departments want to continue to be a part of VIB, they have to perform in terms of patents, revenues, spin-offs, etc.

To keep track of what's happening in the different research groups, there are regular meetings between the labs and TTO officers. At least once every month, the TT officers visit all research groups in person to discuss the progress they are making.

The TTO process starts with a Record of Invention (ROI). This can be a "formal" document, a first draft of a scientific paper, a poster, a set of experimental data with extra background info or a telephone call followed by any of above written documents. After this the ROI is evaluated. The TTO will search the "research finding" for prior art, will evaluate patentability, market and feasibility of the invention, will inquire whether material is used from third parties (e.g. by MTAs) or financing by a non-VIB source (such as IWT, IOF, FWO or EU), will preliminary inquire who the inventors are, will consider drafting a patent application, and, if the previous steps are all positive, will file a patent application with a recognized national (BE), regional (EP) or international (PCT) IP Office.

Inventions are evaluated on four levels: firstly, exclusion criteria are checked to determine who the owner and who the applicant(s) of the invention are. Secondly, the patentability of the invention is looked at, in other words: is it novel, inventive and industrially applicable? Thirdly, the TTO wants to be sure of the commitment of the inventors, because they are needed to further develop the invention. Last, but certainly not least, the commercial value of the invention is taken into consideration. The main questions that arise are: is the invention commercially attractive enough to entice more than one company to license it? And, will you find a company willing to pay for the development of your invention? To address these questions, several other questions should be posed:

- What is the scope of applications? The invention can be a niche application or it can be a technology or product platform; the

invention can be applied in one to many fields or it can be used in one to many applications.
- What is the market situation? The market can be small or large, emerging or saturated, and the product can be an incremental improvement or it can be a disruptive innovation.
- How much freedom to operate is there? The exploitation of the own patent can be dependent on third party patents; third parties holding dominant patent rights can be willing to grant licenses or not.
- Are there competitive (patented) inventions? There are a lot or no competing (patented) inventions; the invention plays in a field with little or many candidate licensees.

The TTO uses several parameters when deciding whether to go for a (service) spin-off route or a licensing route. These parameters assess the following:

- Patent Protection
- Freedom to operate
- Scope of applications
- Market
- Stage of development
- Product type
- Scientists
- Research program

Based on the evaluation of these parameters, 50% of all projects disclosed from 2005–2010 were pursued further by the TTO in terms of patent protection, licensing, start-up formation or other form of commercialization. The stringency of the tech screening process is somewhat appropriate for the organization and surrounding region, although it could be more severe. Sometimes the follow-up of a certain project takes too long and should be stopped in an earlier stage, in order to not waste valuable time and other resources (Leyman, 2012; Staelens, Personal communication, 30 November 2012).

2.2.3. IP management

Patentability, freedom to operate and whitespace are always evaluated before patenting, and in most cases potential licensees are identified before patenting. A commercialization strategy or "technology roadmap" is not always planned before filing. The most important aspects in the patent strategy are timing, geographical scope, patentability and the involvement of the inventors. Also, but less important, are the patent enforceability and the cost containment. Communication with patent offices is not merely important, it's just necessary to file a patent and therefore inherently included in the patent process.

Additionally, the funds needed to follow the IP strategy and to support patent application and enforcement are available at VIB. The money to cover patent costs comes from a licensee, in the cases where there is one. In cases where there isn't a licensee, the institute covers the costs. The investor or public sources are never used to pay these costs.

In the formation of the commercialization strategy, feedback from experts in the industry is not necessarily asked for. If technology is deemed to be at a too early stage, yet has potential, the TTO assists the investigator in finding funding or partners for further development. This further development is then guided by recommendations from industry and the TTO. In the end, the TTO advises and supports scientists in the strategic choice of either licensing out the technology or creating of a company based on the technology.

The generated technology is marketed in several ways:

- TTO contacts potentially interested industry players directly, e.g. by sending flyers to selected companies, by addressing its network of direct contacts or by attending biomeetings and talking directly to the right kind of people.
- The summary of the technology is posted on the institute's website www.vib.be.
- Through the investigator's own contacts in the industry. Sometimes researchers are the best ambassadors or salesman for the technology!
- In 2002 VIB launched the VIB deTECHTor, an electronic (regional) newsletter to make VIB's know-how and tech transfer

activities accessible in a structured manner to companies. It keeps the Flemish biotech industry up-to-date regarding scientific, technological and commercial opportunities at VIB.
- Nowadays, VIB publishes "VIB News", a quarterly printed newsletter for VIB.

Due diligence to evaluate the organizational health of potential licensees is not conducted. On the other hand, the motivations for licensing the technology, as well as the licensee's plan for developing the technology, are made clear from the start, and special measures are taken to ensure that the licensee will put forward the required resources to develop the technology. In this way, the VIB wants to make sure that a company does not take a license just to block a competitor.

In order to speed up negotiating a patent and to lower associated costs, different types of license/option templates are used. However, the office makes no distinction between start-ups or large companies. This means that none of their policies encourage start-ups in particular and that their guidelines are the same whether you are a start-up or a large, established company. There is no deliberate policy either that covers when to license exclusively and when to license non-exclusively. In other words: every case or project is different.

It can be stated that the relevant TTO staff members are aware of different licensing policies and other models of IP transfer, since there are two patent attorneys and four or five business development managers in-house in the TTO team (Leyman, 2012; Staelens, Personal communication, 30 November 2012).

2.2.4. *TTO support of entrepreneurs*

The VIB TTO supports entrepreneurs extensively in several ways: with developing their business plans, in the recruitment of external management, in the composition of a technology platform, in finding office space and equipment and in searching for funding. In accomplishing this last goal, the TTO makes itself familiar with federal grants, foundations, accelerator funds and the like and develops good contacts with the venture capital community and other investors.

2.2.5. Results and achievements in terms of valorization

The technology transfer approach of VIB has proven to be very successful. This is easily demonstrated by its patent and spin-off activities.

2.2.5.1. Basic tech transfer statistics

As patents are included in the performance criteria imposed by the Flemish Government, VIB is developing a strong patent portfolio (Table 2-2). Fifty inventions on average are recorded per year, with an increasing number in recent years (73 in 2011). Of these inventions, approximately 50% are patentable, which results in more than 23 patent filings per year (34 in 2011). In 2012, VIB held 731 granted patents that were combined in 183 patent families. Approximately 61% of these patent families deal with external parties: 71 are licensed to companies, 29 families are transferred (usually to start-up companies) and 12 families are used in strategic deals.

This large success in spin-off creation has some implications on the available office space. The offering of office space, lab space and infrastructure by the VIB is satisfactory. Let it be clear that the quality is excellent, but every new incubator/accelerator building is immediately completely filled with companies, so the demand is still higher than the supply. This space and infrastructure is offered at market price to the companies, there are no discounts whatsoever. One reason for this is that the incubators/accelerators are established as spin-outs from VIB.

Table 2-2: Basic TT statistics of VIB. *Source*: Leyman, 2012.

Total biotech disclosures (2005–2010)	280
New biotech patent applications (2005–2010)	125
Income-generating licenses in biotech	71
Total license revenue for biotech licenses (EUR)	15.000.000* EUR
Spin-offs	12
FTE employed in spin-offs (2012)	540

*(Including License Income, R&D Collaboration Income, Service Income, Income from Sale of Shares, Etc.)

2.3. Implementing best practices in your region

2.3.1. Implications for TTO

- Attract qualified people that are experienced in the various fields of technology transfer.
- Develop and use a well-structured screening and assessment process to select the most promising inventions.

2.3.2. Implications for universities/research organizations

- Establish or further develop top quality biotech research groups within the university.

2.3.3. Implications for policy makers

- Provide a dedicated policy and structural funding to set up an organization like VIB.
- Make valorization oriented parameters part of the periodic evaluation.
- One prerequisite, however, to set up an organization like VIB is to have existing excellent biotech groups in the region.
- Provide financial support to build-up TTO organizations.

2.4. Future opportunities

Currently, we witness that bio-incubator space is immediately occupied by new biotechnology firms. This indicates that there is a vibrant start-up scene in the region and that increasing the available incubator space may offer opportunities to those start-ups presently underserviced or looking for lab space. Due to the structure and working principle of VIB, there is an inherent pressure on academic staff to perform according to the predefined criteria that are imposed by the Flemish Government. There might be untapped resources, knowledge or opportunities present within these research groups that can be unlocked by providing more (financial) means and incentives for testing new ideas or keeping an idea secret for a longer time period.

2.5. Best practices

- Long-term government support: the Flemish government has supported the VIB since its official founding in 1996. Prior to this, the Flemish authorities also initiated other programs to support (biotechnological) science valorization, such as the VLAB and the IWT.
- VIB structure and financing model:
    - Evaluation every five years by the Flemish government, based on performance criteria, to "earn" the next investment of the government. This makes sure that everybody is after the same outcome with no differing priorities.
    - Reward system: the better a research group performs, the higher its share of the following governmental investment.
    - VIB has the freedom to choose the way in which it aims to reach the targets that are set by the Flemish Government.
- VIB has a large and experienced technology transfer team: 12 real officers, each with their specialization and all dedicated to biotech
    - The TTO of the VIB is a centralized office, located in the headquarters of the institute. The VIB TTO employs 16 people, of which 12 are actual officers and four are administrative staff members. The 12 officers are composed of four IP Managers, of which two are patent attorneys, four licensing managers, two business developers, one contract manager and one general manager, who is also involved in business management.
    - TTO is centralized and serves biotech research groups in different universities. These groups are mainly part of VIB, but also some non-VIB groups are supported.
- Clear selection criteria of technologies:
    - Patentability of the invention.
    - Commitment of the inventors.
    - Commercial value of the invention.
    - Will there be more than one company willing to take a license?
    - Will you find a company willing to pay for the development of your invention?

- What is the scope of applications? Niche or platform; one or many fields; one or many applications.
- What is the market situation? Small or large; emerging or saturated; incremental or disruptive innovation.
- How much freedom to operate is there?
- Are there competitive (patented) inventions? Little or many candidate licensees?

2.6. References

Papers, reports, books, conference presentations and websites

D'Hondt, K. (2012). *Science and Innovation Policy Measures: Flanders & Biotech* [Powerpoint Presentation]. ETTBio site visit, Brussels, 10.10.2012.

Leyman, B. (2012). *Valorisation of Research*: VIB. ETTBio site visit in Brussels, 11.10.2012.

VIB (2012a). History of VIB. Retrieved from: http://www.vib.be/en/about-vib/organization/Pages/History.aspx. Accessed on 04.06.2013.

VIB (2012b). VIB renews General Management — Interview with Rudy Dekeyser. Retrieved from: http://www.vib.be/en/news/Pages/VIB-renews-General-Management---Interview-with-Rudy-Dekeyser.aspx. Accessed on 06.06.2013.

Interviews

- 30.11.2012, Jan Staelens, VIB.
- 26.11.2013, Kathleen D'Hondt, OECD Science and Technology — EWI.

3

CASE STUDY 3: The Creation of a New Technology Transfer Office

Magdalena Powierża and Piotr Potepa
(*International Institute of Molecular
and Cell Biology, Poland*)

3.1. Setting the scene

3.1.1. *Introduction to IIMCB and Ochota Biocentre*

The International Institute of Molecular and Cell Biology (IIMCB) is made up of nine research groups, including a joint one with the Max Planck Institute of Molecular Cell Biology and Genetics in Dresden. The focus of the research is on fundamental biomedical issues. The Institute is financed in part from the national budget and in part from other sources (Ministry of Science and Higher Education, Foundation for Polish Science, National Science Center, National Center for Research and Development, Framework Programs of EU, Max Planck Society, Howard Hughes Medical Institute, European Molecular Biology Organisation, National Institutes of Health, Wellcome Trust, European Science Foundation, etc.). About 80% of funds arrive as competitive grant awards received by the group leaders. In the recent evaluation of scientific institutions in Poland, IIMCB was ranked first among all institutions in the field of biological sciences. IIMCB has ten faculty members and 114 research employees. All ten faculty members and 114 research employees are active in biotechnology-related fields (IIMCB, 2014).

IIMCB belongs to a research consortium called Ochota Biocenter. These are altogether five bio-med institutes and one technical. The total amount of researchers in the campus is approximately 1,500, including approximately 400 PhD students (Ochota Biocentre, 2014).

3.1.2. General information about the BioTech-IP

TTO-BioTech-IP is a young Technology Transfer Office established at IIMCB in March 2010. It serves the purposes of the Ochota Biocenter consortium (medical devices, therapies, biotechnology and therapeutics) with the aim to offer the services to other research entities. BioTech-IP is 100% self-funded — it relies on the money secured in grants for various aspects of technology transfer, which means that the bulk of their responsibility is strictly related to the execution of these projects. There is no internal fund for unexpected activities that are not covered by any of the BioTech-IP projects. It leads to a situation in which core tech transfer activities are sometimes done in an indirect way and outside the budget. The mission of the TTO is to raise awareness with scientists towards applied science and to assist in the commercialization of mature projects. The immediate goal of BioTech-IP was to encourage scientists to cooperate and get the technology transfer mechanism going. In 2013 BioTech-IP was granted government funds to develop a special purpose vehicle entity to facilitate technology transfer of publically owned technologies of Polish universities and institutes to private companies. BioTech-IP has also applied for its own seed money from EU funds in order to be able to invest in proof-of-concept trials and to establish spin-off companies. There are nine people altogether working part time for TTO-BioTech-IP: four are employed directly by IIMCB while different research institutes employ five scouts (BioTech-IP, 2014).

3.1.3. History of technology transfer in Poland

The history of technology transfer in Poland is not very long. The first TTOs were established in the late nineties at the biggest technical universities. Between 2007 and 2013 there was a clear rise in the number of TTOs along with a considerable rise in the amount of money devoted to this by the EU structural funds and governmental sources. This is also the time when serious debate started about the future of technology transfer (debate that continues up to the present day), which was followed by improvements in the legal system.

3.2. Identifying best practices

Biotech-IP was created to encourage scientists to undertake applied research projects and support them in the commercialization of results of their work. The TTO was founded because there existed a huge intellectual potential with Ochota Biocenter researchers while an institution supporting technology transfer was missing. Building the foundation of the TTO has been a long process because first scientists must get to know why they should follow the technology transfer route and then how to do it. They also need to know that TTO staff are on their side, working on behalf of the institute and not for the industry. It was not obvious at the beginning that it would be necessary to establish trust between the parties. In the next part we focus on the TTO history, present activities and plans for the future, identifying best practices along the way.

3.2.1. BioTech-IP's history

The idea to create a TTO within the International Institute of Molecular and Cell Biology was a bottom-up initiative of three employees — a person working in the International Cooperation unit with a vast experience in FP7 project management and two PhD scientists engaged at that moment in various tech transfer initiatives. In 2009 they realized that there had been no such thing as tech transfer awareness among the scientists, let alone tech transfer mechanisms at IIMCB or other Ochota Biocentre institutes. They prepared an action plan in which they outlined the structure of a future TTO and looked for sources of funding. They presented their idea to the

Director of IIMCB who supported this idea. At this stage the main indispensable elements of the plan were:

1. Enthusiasm of young (and inexperienced) people who saw an opportunity to explore a *Terra incognita* of tech transfer and wanted to create a new unit at the campus.
2. Full support of the IIMCB's director who noticed the need to create a TTO and trusted these young people and let them act.
3. Favorable institutional environment in Poland, which supported such initiatives financially. A TTO could have only been brought to life if external funds had been secured as the IIMCB could not invest.
4. An idea to serve the needs of six institutes of Ochota Biocentre instead of just IIMCB.

In 2009 the three founders of the TTO wrote their first grant applications — "Innovativeness Creator" — and in 2010 the grant contract was signed for approximately 100,000 euros for employment, equipment and computers in a TTO office, training courses for the staff and scientists in IPR and tech transfer. A best practice at this stage was to elicit representatives from other Ochota Biocenter institutes who enjoyed greater trust from local scientists and thus could work as scouts. Thanks to this grant, the TTO staff could verify grant applications and publications prepared in all the institutes *vis a vis* patentability of the results contained therein. The scouts are in contact with scientists and report the developments in their work. Scientists also receive BioTech-IP's help in applying for funds devoted to applied research grants such as OP IE 1.3.1 and 1.3.2 and Applied Research Programme (BioTech-IP, 2014).

In 2010 TTO signed another contract for approximately 380,000 euros in order to organize:

- Scholarship program for PhD students
- Internships for scientists at companies
- Training courses on tech transfer, IPR and soft skills.

These two projects provided the foundations for Bio & Technology Innovations Platform (BioTech-IP) which was formally set up in March 2010.

3.2.2. TTO team and structure

At present, BioTech-IP is part of IIMCB and is subordinate directly to IIMCB's director. In October 2012, BioTech-IP received a new office in the IIMCB's building (approximately 18 square meters), fully furnished and equipped with the support of IIMCB and BioTech-IP's grant money. The TTO is independent in its decisions, which however must be approved by the IIMCB's director.

The TTOs staff is composed as follows: two PhDs with backgrounds in biotechnology and tech transfer, one PhD student in life sciences (all of them part-time), two administrative employees with background in project management (full time). Additionally, there are five scouts who represent the Ochota Biocenter institutes in BioTech-IP. There is one outsourced tech transfer manager with experience in the UK and several lawyers and patent attorneys.

3.2.3. TTO tasks and responsibilities

BioTech-IP is a young TTO created to establish technology transfer mechanisms in Ochota Biocentre institutes as an internal initiative of IIMCB employees. First they had to gather funding for their activities and teach scientists how to behave in the reality of tech transfer. Thus the TTO activities are strictly related to the projects they are financed from. From the beginning, a vast part of the activities was training for scientists on IPR, tech transfer, negotiations, preparation of R&D projects, communication and presentation of R&D results to business partners, etc. Students are not provided with such information during regular studies thus BioTech-IP's courses fill in the educational gap. The trainings have been welcomed with great interest and were evaluated highly by the participants. So far, approximately 650 scientists have been trained.

The TTO assists the scientists in the following:

- Applying for public grants for patent protection — approximately one million Euro have been raised with the help of BioTech-IP for seven technologies developed at the campus.

- Applying for public funds which initiate collaboration with industry in a joint project consortium.
- Finding business partners via: the BioTech-IP's website which contains information of promising technologies, regular science-to-business brunches during which scientists present their projects to invited business partners, brochures on Ochota Biocentre services and projects.
- Preparation of legal and financial issues of technology transfer.

The TTO organizes two big training programs:

- For PhD students who are carrying out applied research projects. Students are supported with BioTech-IP's grants of up to 750 euros a month.
- Internships at companies for scientists and PhD students — up to 500 euros a month on top of their regular salary.

3.2.4. *Financing*

Each aspect of BioTech-IP's activity is financed from grant money, provided that it has been planned in a given project. This means that BioTech-IP is limited in their actions if the budget does not allow expenditures in unexpected cases. If an IIMCB technology appears and there is no grant money to support it, IIMCB pays for it. However, cases of technologies from other affiliated research institutes are usually more problematic. At present, BioTech-IP has a budget of more than 1.5 million euros to cover all its expenses foreseen in the projects' budgets for the period of April 2010–May 2015 (BioTech-IP, 2014).

3.2.5. *Achievements*

Thanks to the creation of TTO, since March 2010 the following results have been achieved:

- Approximately 650 scientists from the whole campus have been trained in such areas as: intellectual property protection and

management, project and team management, negotiations with business partners and financial and legal aspects of tech transfer.
- Thirty PhD students from the whole campus received scholarships to execute applied research projects. They results are monitored and the projects are further developed by BioTech-IP.
- Six scientists from the whole campus have completed up to two month paid internship at biotech companies and got involved in subsequent cooperation with these companies.
- Seven PCT patent applications were submitted only from the host institute IIMCB (before there had been one).
- One spin-off is under creation based on one of the IIMCB technology.

3.2.6. Future plans

BioTech-IP plans to expand its activities and resources and form a special purpose vehicle (SPV) together with IIMCB. The focus of the new company will be strictly on IPR management, technology evaluation and commercialization of technologies that originate at the campus through licenses and creation of spin-off companies. With this in mind, BioTech-IP applied to a special governmental program and received one year's funding in which it is to prepare an action plan of this venture along with all necessary legal and tax solutions to be implement afterwards in phase B, i.e. factual creation of a SPV.

3.3. Implementing best practices in your region

As mentioned before, this case study presents the creation and first three years of existence of a new TTO. To be a mature and well-functioning entity it needs more time and experience not only on behalf of a TTO but also other stakeholders such as policy makers and the research organization itself.

3.3.1. Implications for TTO

The vast majority of present BioTech-IP staff is a PhD or at least has a master's degree in biotechnology. The officers need to be able to communicate with the scientists and easily understand the technology on which they are working. However, at the same time they also need to be aware of the market opportunities and preferably have experience in commercializing these types of technologies.

Attracting these types of persons is definitely a challenge for the TTO. It is difficult to find staff that combines these two capabilities — understanding the content matter of the biotechnology research and the market reality with the view on finance and legal aspects. It is also important to have staff members who understand the implications of the Polish institutional setting. Therefore, employing an international expert with experience in tech transfer yet with little understanding of the Polish systems is not a solution. Luckily, the subject of technology transfer is attractive and more and more scientists seek to advance their career via technology transfer. In their everyday work, Biotech-IP has a chance to select people that may be interested in long-term collaboration: scientists interested in working with the TTO, a legal adviser or a freelancer. The key selection criterion is passion and involvement in this type of work.

3.3.2. Implications for research organizations

IIMCB participates in a small fraction of the BioTech-IP's expenditures, which made the TTO staff look for the ways of securing their own budget. IIMCB provides office space and equipment, as well as administrative support and, as previously mentioned, covers unexpected expenditures. What is most important is the "freedom to operate" that BioTech-IP had in planning its activities. The mutual relations between IIMCB and Biotech-IP are based on trust, respect and symbiosis.

3.3.3. Implications for policy makers

Policy makers are perhaps the most important stakeholders in the whole process. They create the legal framework of technology transfer.

Nowadays in Poland there is a heated debate over the regulations in this respect among different ministries setting foundation for a reform of the system. One of the institutions supporting science — National Center for Research and Development — launched the program SpinTech, whose aim is to develop different schemes of founding spin-offs at public research institutions by a SPV. The outcome of this program is to come up with practically verified "recipes" for establishing spin-offs at universities or institutes from legal and taxation points of view. Biotech-IP participates in this program (NCRD, 2014).

3.4. Future opportunities

- Deployment of mechanisms of investment into spin-offs. Preparation of legal, tax and financial framework of such undertakings.
- Establishing own investment fund which is *sine qua non* of further development which would support creation of spin-offs at the campus.
- Tightening collaboration with experts who understand the Polish systems.

3.5. Best practices

- Motivated team who gave raise to establishment of the TTO.
- Availability of public funding which supported the creation of a TTO.
- Well-defined action plan and future steps to be taken in order to expand into a mature phase.
- Good relations with the scientific environment — BioTech-IP staff is seen as work mates rather outside businesspeople.
- Expanding the pool of experts who collaborate with BioTech-IP — legal, tax and financial advisors, patent attorneys.
- Creating a new reality of tech transfer in Poland together — BioTech-IP, their advisors and scientists work out the most efficient mechanisms in the current circumstances.

3.6. References

Papers, reports, books, conference presentations and websites

Bio & Technology Innovations Platform (BioTech-IP) (2014). Retrieved from: http://www.biotech-ip.pl. Accessed on 04.07.2014.

International Institute of Molecular and Cell Biology (IIMCB) (2014). Retrieved from: http://www.iimcb.gov.pl. Accessed on 04.07.2014.

National Center for Research and Development (NCRD) (2014). Retrieved from: http://www.ncbir.pl. Accessed on 04.07.2014.

Ochota Biocenter (2014). Retrieved from http://www.biocentrumochota.pan.pl. Accessed on 04.07.2014.

Further reading

According to the Mazowieckie Biuro Planowania Regionalnego w Warszawie (2013) "Strategia Rozwoju Województwa Mazowieckiego do 2030 roku; Innowacyjne Mazowsze". Retrieved from: http://www.mbpr.pl/user_uploads/image/PRAWE_MENU/PROJEKT%20STRATEGII/SRWMPROJ.pdf. Accessed on 04.07.2014.

4

CASE STUDY 4: A Model for IP Transfer and Shareholding for University Spin-Offs: The "Dresden Model"

Alexander Funkner, Nadine Schmieder-Galfe and Oliver Uecke
(Technische Universität Dresden, Germany)

4.1. Setting the scene

The Technische Universität Dresden (TUD) has its roots in the Royal Saxon Technical School, which was founded in 1828. TUD covers a wide range of courses of study such as Engineering, Mathematics and Sciences, Humanities, Social Sciences as well as Medicine.

Together with the university hospital, TUD is a major healthcare provider. TUD has the first DFG Research Center (Center for Regenerative Therapies), which is the first East German cluster of excellence to be supported by the Federal Government within the scope of its Initiative of Excellence. This initiative supports excellent research at German universities in order to in order to achieve a world-class position in science (Excellence Initiative, 2014). TUD has become the federal elite university with its "DRESDEN-concept" (Dresden Research and Education Synergies for the Development of Excellence and Novelty). The aim of the DRESDEN-concept is to support Dresden's leading scientific areas, thereby using the synergies of all partners in terms of research, education, infrastructure and administration (see also Case Study 17). Within this concept, TUD also strongly encourages entrepreneurship and technology transfer. The TUD has 3,651 faculty members and 2,261 research employees. Additionally, 475 faculty members and 300 research employees are active in biotech-related fields. TUD has around 35,900 undergraduates and 5,900 graduates, with 2,200 undergraduates and 470 graduates in biotech-related fields. TUD has also a business faculty which employs 109 faculty members, has 1,319 MBA students and 112 PhD students (TU Dresden, 2012).

TUD belongs to the leading universities regarding patenting in Germany. Annually, 100 inventions are being filed as a patent application — one-third are international applications and approximately one-quarter of the applications are in the field of life sciences. Fifteen licenses have been granted and fifty patents have been sold to regional SMEs as well as international companies. Additionally, patents that are held by the TUD are an important factor to support the formation of new companies.

The following players are all involved in technology transfer at TUD. The Department for Research Promotion and Transfer (DRPT) within TUD is responsible for IP protection, R&D-related contracts and coordinates the commercialization of R&D results with external partners. As a transfer point of the TUD, the DRPT is the contact institution for questions concerning knowledge and technology transfer, offering a broad range of technology transfer services to the employees of the university. The DRPT is working at the interface between science and industry in the field of patent application and exploitation. With its Patent Information Centre the Department proactively supports researchers addressing all related patent topics.

The entrepreneurship initiative dresden|exists at the TUD was founded in 1999 with the aim to motivate, qualify and coach students, graduates and researchers in the field of entrepreneurship. It supervises 50–70 start-up ideas each year, which results in 20 start-up companies every year. dresden|exists is connected to the Chair of Entrepreneurship and Innovation (Professor Michael Schefczyk), with teaching and research activities in the field of venture capital, financing of spin-offs, innovation management and technology transfer in biotechnology. dresden|exists supports future founders from the first idea until the company is founded. dresden|exists offers its services also to research institutes in Dresden: Leibniz Association, Max Planck Society, Fraunhofer and Helmholtz. dresden|exists employs three people dealing with biotech-related entrepreneurs. One has a business PhD, and the two others have industry experience in a biotech start-up and pharma company.

GWT-TUD GmbH (GWT) is a service provider for knowledge and technology transfer. GWT commercializes research from the TUD, the University Hospital and other research institutions. It links industry and public research institutes in the Free State of Saxony. GWT is a medium-sized, private enterprise and is a 100% subsidiary of the TUDAG (TU DRESDEN AG). The company has been in the market for 15 years with annual sales of 27 million euros in 2013 (GWT, 2014). It is one of the largest technology transfer companies in Germany. Customers in Saxony, across Germany and within international projects are serviced by locations of GWT in Dresden, Chemnitz and Berlin. There is a total workforce of 20 full-time employees working in the biotech sector at GWT. About one-third of these employees has a biotech-related university education, all have industry experience and most have a PhD. One of the employees is a lawyer.

Biosaxony is the biotech cluster organization serving as a coordinator for the biotech sector as well as a lobbying on a political decision-making level with key activities such as representing the Saxony biotech sector regionally and globally, communicating between political, scientific and economic interest groups, supporting start-ups and growth companies as well as attracting new companies to the region.

In 1994 TUD was setting up a new framework for technology transfer. At this time, professors and research and teaching assistants at German universities were allowed to decide by themselves to file a patent or not and to commercialize their inventions, a right called the university teachers' privilege. TUD started to re-establish patent offices, which had already existed before 1989 in the former East Germany (GDR). In return for filing the patents, TUD offered the inventor the possibility of support during commercialization. As a result, TUD was the first university announcing its goal to strengthen Saxony as a business location and to sharpen Dresden's profile as a high-tech location. In 2002, the reform of the university teachers' privilege became effective in the whole of Germany. German universities became owners of the IP of their employees and were then legally able to claim and commercialize the IP (as it was already the situation for inventions in companies). TUDAG (TU Dresden Aktiengesellschaft) was then founded in order to conduct technology transfer and to create university spin-offs.

4.2. Identifying best practices

In 2000 the Society of Friends and Sponsors of the TU Dresden e.V. decided to found the TUDAG. It is a private company that fulfills the task of technology transfer for TUD. The aim is also to strengthen the economy of the Free State of Saxony. The Society of Friends and Sponsors is the only shareholder of TUDAG. The only link between TUD and TUDAG is given through staff active within both organizations. TUDAG can perform certain tasks where the university is limited in how involved it can be. These tasks are for instance conducting clinical research studies, acquiring R&D contracts with industry partners and also founding companies. Since the beginning, TUDAG has been very successful with its business model and is economically independent (Dradio, 2012). In 2012 its total revenue was 51.2 million euros with a profit (EBIT) of 4.2 million euros (TUDAG, 2013).

Benefiting from profit of the TUDAG holding, TUD will receive funding indirectly through the Society of Friends and Sponsors of the TU Dresden e.V. In addition to that, TUDAG is also active in fundraising, acquiring R&D contracts with industry partners and investing in research equipment as well as financing scientific and technical staff. As a holding, TUDAG has shares in 15 companies, dealing directly with either technology transfer or with educational tasks. TUDAG as a transfer player links R&D competences of the TUD with transfer activities including clinical studies, industry projects, educational courses or spin-offs.

Derived from the need to create an easy access to university IP for TUD spin-offs, a unique shareholding model has been developed; the so-called "Dresden Model". In this model, TUD, TUDAG and the new spin-off cooperate in a trilateral way in order to enable researchers to start up a business based on university IP. Firstly, once a researcher has made an invention, TUD claims the invention and files a patent. Secondly, if this researcher or a team around the researcher founds a company, the patents are transferred to this company via the TUDAG. The patent will actually be sold to the company for a price that covers all actual patent related costs plus overhead. Thirdly, the new company will additionally get a right of first offer for related technologies developed in the future at the university. As a compensation payment for this transfer of intellectual property rights, TUDAG receives shares of the newly founded company.

In the Dresden Model, TUDAG is involved in holding company shares, and not TUD directly. TUDAG is not acting as trustee for TUD. TUDAG is an independent shareholder. There are several good reasons: as a medium-sized enterprise, TUDAG first of all is much more flexible compared to a university which is economically tightly restricted as a subject of rights and duties facing Saxon Ministry for Science and Art (Sächsisches Ministerium für Wissenschaft und Kunst) and its obligation regarding the Saxon Court of Audit (Sächsischer Rechnungshof) (Rasmussen-Bonne, 2009). Of note, if TUD would be directly involved in a university spin-off company as a shareholder, the company would also become a liable subject facing both the legal and financial rules of the Saxon Ministry for Science and Art and the Saxon Court of Audit. Furthermore, TUDAG as a shareholder can also act independently of external influences regarding all kinds of management decisions including human resources in the management board or growth decisions of a company by capital increase instead of profit distribution to shareholders. As TUDAG is also active in fundraising, it is also much easier for TUDAG to invest in second or third round financing compared to TUD. In an exit case when the TUD spin-off is going public or acquired, exit revenues for the TUDAG stake are shared 50/50 between TUD and TUDAG. TUD does not refund TUDAG for any losses derived from the shareholding nor does it pay a management fee for managing the shares.

As a result, this model can be seen — not only under legal circumstances — as a best practice to enable and support technology transfer from science towards university spin-offs.

4.2.1. Evolution of the "Dresden Model"

Since 1957, professors, research and teaching assistants at German universities were allowed to file patents and commercialize their know-how and inventions (intellectual property). They were the owner of this intellectual property. A negative aspect was that they had to cover relating costs, e.g. for patenting. In 1994, based on past experiences, TUD decided to use the former concept of claiming patents via patent offices, which was a successful concept already during GDR times. In return for filing the patent, TUD offered support to inventors while filing and exploiting these patents as well as covering all related patent costs. Invention valorization was in line with strengthening Saxony as an economic region and profiling Dresden as a high-tech location, a declared objective by TUD.

As mentioned previously, in 2002 the German government decided to change the employee invention act and abolish the university teachers' privilege. As was already the case for companies, universities now also received ownership of any intellectual property created by their employees. The German government's objective in doing this was to stimulate knowledge and technology transfer. TUD acted before this legal change happened, and thus, it had a pioneering position with a unique model for tech transfer in Germany.

4.2.2. The Dresden Model today

To date, the Dresden model has been successfully applied for 21 technology-oriented patent based ventures. One of those ventures is the Dresden-based company Riboxx Pharmaceuticals (General Management TUDAG, 2013). It was founded in 2010 by Dr. Jacques Rohayem, a former group leader at the Medical Faculty of TUD.

Riboxx is active in the field of nucleic acids engineering and bio-processing, thereby developing and validating novel molecules for *in vivo* applications. Recently, one of the TUD start-ups (Novaled) was acquired by Samsung, valuing Novaled at 260 million euros. At this exit stage, TUDAG was still a shareholder (Novaled, 2013).

4.3. Implementing best practices in your region

To implement the Dresden Model in your region requires actions on university and research institute level as well as on policy level. Policy maker should set up a legal framework, which ensures universities can hold shares in their spin-offs, directly or indirectly via a model such as the Dresden model. The university and research organizations should have a clear strategy for technology transfer and spin-off creation. University spin-offs should get the possibility to get access to past and future university IP. There should be an efficient and transparent process for technology transfer ("time is money"). In addition, the university and research organization and its TTO should provide support along the whole technology transfer — and entrepreneurial process.

4.4. Future opportunities

The Dresden Model could potentially be extended for licensing when no spin-off is created but a technology is licensed to industry.

4.5. Best practices

The Dresden Model is a framework, which ensures the following:

- The possibility to create university spin-offs in order to commercialize university IP.
- The university participates in the increase of the firm value and profits of the spin-off through holding shares (indirectly via a third party).
- The spin-off easily can access past and future university IP.

4.6. References

<u>Papers, reports, books, conference presentations and websites</u>

Dradio (2012). TUDAG — Wissenschaft, die Geld einbringt. Retrieved from: http://www.dradio.de/dlf/sendungen/campus/1712071/. Accessed on 12.07.2014.

Excellence Initiative (2014). Retrieved from: http://www.excellence-initiative.com/start. Accessed on 12.07.2014.

GWT (2014). Retrieved from: http://gwtonline.de/gwt/daten-fakten/. Accessed on 12.07.2014.

Novaled (2013). Retrieved from: www.novaled.com. Accessed on 01.10.2013.

Rasmussen-Bonne, H.-E. (2009). Technologietransfer an der TU Dresden. *Biotechnologie*, 9, 2009.

TU Dresden (2012) Statistik Immatrikulationsamt.

TU Dresden (2013). Retrieved from: www.tu-dresden.de. Accessed on 12.07.2014.

TUDAG (2013). Retrieved from: http://www.tudag.de/die-tudag/tudag-in-zahlen/. Accessed on 30.06.2013.

<u>Case study interview</u>

- 2013, General Management, TUDAG, Dresden.

SECTION 2: FUNDING

Introduction

A well-structured Technology Transfer Office (TTO) that has the experience and expertise to turn raw and promising ideas into successful license agreements or business cases is crucial for the success of technology transfer. The TTO needs to develop a technology transfer process that screens and identifies opportunities, supports inventors in protecting their intellectual property and commercializing their inventions either by setting up a company or concluding license agreements. However, to bring together an experienced technology transfer team, support IP strategies and commercialization, one needs to have financial resources. In this section, we focus on (1) how TTOs finance their activities and (2) funding mechanisms and programs necessary to stimulate and support technology transfer. We look at funding mechanisms on organizational, regional, national and international levels that facilitate the transfer of know-how and technologies from academia to industry.

Universities and TTOs face a number of resource constraints in transferring knowledge to society. Universities in the UK and continental Europe indicated that access to finance — with access to venture capital figuring highly — is one of the main factors impeding the technology transfer and more precisely the creation of university spin-out companies (Wright *et al.*, 2006). The existence of a funding gap between the demand for finance from entrepreneurs/inventors and the willingness of suppliers to provide this finance has long been recognized (Shane, 2004; European Commission, 2000). Venture capital firms' provision of risk capital has been seen as a major solution to bridge the so-called equity gap for inventors that want to commercialize their intellectual property. However, venture capital firms in Europe especially have traditionally been criticized for being reluctant to invest in early stage high-tech investment (Lockett et al., 2002). Research in the UK concluded that the majority of venture capital investors prefer to invest in spin-offs after the seed stage, particularly once proof of concept has been achieved. The finding does not seem to be unique to the UK as Moray & Clarysse (2005) showed that less than one out of ten spin-offs could attract venture capital. These studies reported that although universities point to the difficulties in attracting venture capital and business angel investment, they seldom develop mechanisms to increase the possibility of internal funding and debt financing as an intermediate step towards venture capital investment (Wright et al., 2007). The fact is that inventors often do not have the ability to attract the financial resources to develop a prototype or investigate the technological feasibility and target market needs. This means that the funding gap discussion, which is still very lively and prevailing, forces universities and government to find solutions to deal with this gap.

In what follows, we show how TTOs from different parts of Europe deal with this funding gap, managed to finance their technology transfer activities and offer and use different funding mechanisms to support spin-outs. There are different approaches to deal with this funding gap that can be used in combination with one another. A way to approach this funding gap is that universities and more specifically TTOs, act as a sort of venture capitalist that uses less strict milestones to support spin-offs to bridge the funding gap in the early stages. The TTO manages their own investment fund and in some cases also participates in investment rounds beyond the early stages of a spin-off. In Case Study 5, we see that Imperial Innovations has a technology transfer team and a ventures team. The ventures team contains a lot of prior investment experience and invests in different opportunities arising from Imperial College and also from Cambridge, Oxford and University College London (UCL). Investments are made from the balance sheet of Imperial Innovations. Moreover, we also focus on how Imperial Innovations uses its ventures team to develop strong links with charity organizations and the venture capital community in London and convince external investors to look at investment opportunities in their university.

Second, governments can offer financial tools to TTOs and researchers to facilitate and financially support technology transfer. Case study 6 explains how the Flemish government set up industrial research funds to support universities to valorize their research. The program gives a lot of freedom to the universities, supports researchers to move from proof of principle to proof of concept and increased the awareness of valorization among Flemish universities. In Case Study 7 we show how the innovation voucher stimulated collaborations between private companies and universities and helped to eliminate the initial distrust between universities and private companies in Czech Republic. In Poland (Case Study 8), European and national funding is managed by the National Centre for Research and Development to support various important activities in the technology transfer process. Different programs are explained that provides financial resources for patent protection, public R&D projects, R&D centers and special purpose vehicles that act as intermediaries between research institutes and the market.

To summarize, all the cases studies below underline the importance of a strong government that considers innovation and entrepreneurship as key drivers of economic recovery and growth. We also noticed that governments in the case studies pay a lot more attention to how and if academics provide evidence of societal impact when they allocate their financial support. Understanding how engagement results in such benefits and simultaneously maintain scientific quality appears now of greater relevance for governments (Perkman *et al.*, 2013). In addition, the university board needs to have a clear vision and focus on technology transfer so that they make use of government funds. Moreover, they benefit from developing strong links with the financial community in their region.

References

European Commission (2000). *Progress Report on the Risk Capital Action Plan*. European Commission, Brussels and Luxembourg.

Moray, N. & Clarysse, B. (2005). Institutional change and resource endowments to science-based entrepreneurial firms. *Research Policy*, 34(7), 1010–1027.

Perkmann, M., Tartari, V., McKelvey, M., Autio, E., Broström, A., D'Este, P., Fini, R., Geuna, A., Grimaldi, R., Hughes, A., Krabel, S., Kitson, M., Llerena, P., Lissoni, F., Salter, A. & Sobrero, M. (2013). Academic engagement and commercialisation: A review of the literature on university–industry relations. *Research Policy*, 42(2), 423–442.

Shane, S.A. (2004). *Academic Entrepreneurship: University Spinoffs and Wealth Creation*. Edward Elgar Publishing, Cheltenham, UK.

Wright, M., Lockett, A., Clarysse, B. & Binks, M. (2006). University spin-out companies and venture capital. *Research Policy*, 35(4), 481–501.

Wright, M., Clarysse, B., Mustar, P. & Lockett, A. (2007). *Academic Entrepreneurship in Europe*. Edward Elgar Publishing, Cheltenham, UK.

5

CASE STUDY 5: Environmental Success Factors of Imperial College's TTO

Robin De Cock
(*Ghent University, Belgium; Imperial College London, UK*)

5.1. Setting the scene

London is a leading global city, with strengths in the arts, commerce, education, entertainment, fashion, finance, healthcare, media, professional services, research and development, tourism and transport all contributing to its prominence. It is the world's leading financial center alongside New York City and has the sixth largest metropolitan area GDP in the world depending on measurement. London's 40 universities form the largest concentration of higher education in Europe. London's largest industry is finance, and its financial exports make it a large contributor to the UK's balance of payments. London has over 480 overseas banks, more than any other city in the world. The City of London is home to the Bank of England, London Stock Exchange, and Lloyd's of London insurance market (City of London, 2013).

Imperial College London is a science-based institution with a reputation for excellence in teaching and research. The Imperial College Business School exists as an academic unit outside of the faculty structure. Imperial Innovations was founded in 1986 as the technology transfer office for Imperial College London, to protect and exploit commercial opportunities arising from research undertaken at the College. In 1997, the Group became a wholly owned subsidiary of Imperial College London and in 2006 was registered on the Alternative Investment Market of the London Stock Exchange, becoming the first UK University commercialization company to do so.

Nowadays, Innovations can be seen as a technology commercialization company that has a Technology Pipeline Agreement with Imperial College London that extends until 2020, under which it acts as the Technology Transfer Office (TTO) for Imperial College London. The Group also acts as the TTO for Imperial College NHS, Royal Brompton & Harefield, Chelsea and Westminster Foundation and North West London Hospital Trusts (Imperial Innovations, 2013).

However, Imperial Innovations was not built in a day and evolved over the years to what it is now. It all started in 1985. There was a complete change in the ownership of patents on work done in universities when Mrs. Thatcher's government in the UK decreed that work done in universities funded by the government (which means most research) should belong to the university, as long as they set up a mechanism to exploit it (Penfold, 2012). One year later, Imperial Innovations was founded as the TTO for Imperial College London to protect and exploit commercial opportunities arising from research undertaken at the College. In the beginning Innovations was a Pro-Rector's office and included just two or three staff. From 1987–1997 the TTO grew very slowly and was mainly focused on IP management (Penfold, 2012).

In 1997, the government launched several initiatives to stimulate technology transfer. One of these initiatives is the UCSF (University Challenge Seed Fund) that provided universities with capital to support their start-ups with small amounts of seed money. Imperial Innovations now had the capacity to offer financial and incubation support and became a more commercial organization. The TTO team grew to a team of 20 persons and became a wholly owned subsidiary of Imperial College London. In 2003, the government again stimulated the growth of TTOs around the country with the HEIF (Higher Education Innovation Fund) initiative. Imperial college enhanced the incubation support and set up the corporate partnerships function (Penfold, 2012).

In 2004–2005, the college developed a new College Endowment Strategy as a reply to the changing environment where UK venture capitalists had moved onto later stages and had fewer funds invested in early university-originated start-ups. As such, much substantial value that was created at the College went unexploited. One of the main successes of this new strategy was the private placement of shares, which resulted in £10 million for the College and £10 million for Imperial Innovations. In July 2006, Innovations completed its Initial Public Offering and was listed on the AIM of the London Stock Exchange as Imperial Innovations Group plc. As a result of the IPO, the Group raised £26 million (+£30 million in November 2007) to invest in its portfolio of companies based on Imperial College London technology. Between 2008 and 2010, the group achieved two important exits. As part of this fundraising, it commenced investment in intellectual property developed at or associated with the universities of Cambridge and Oxford and University College London (UCL), in addition to Imperial College London. The College still holds 30% of the shares. The last couple of years, Innovations invested substantial amounts of money in their portfolio companies (Imperial Innovations, 2013).

Looking back at the history of the TTO, we can conclude that the government, the financial environment in London, the board of directors of Imperial College and the world class research at Imperial College had an important impact on the growth of Imperial Innovations.

5.2. Identifying best practices

In what follows, we explain best practices and environmental success factors at two levels: university level and regional level. At the regional level, we identified the financial community and the government as important factors in the growth of Imperial Innovations.

5.2.1. University level

Imperial College London was one of the first universities which focused on technology transfer and on creating a socioeconomic impact. For decades, these words are incorporated in the mission statement, goals, vision but also in the minds of everyone who works at Imperial College London. Imperial College has a strong history of technology transfer and encourages connections with the industry. Moreover, Imperial College London is one of the only UK universities which has had the application of its research to industry, commerce and healthcare central to its mission since its foundation. The mission statement clearly shows the important relationship with the industry:

> *Imperial College embodies and delivers world class scholarship, education and research in science, engineering, medicine and business, with particular regard to their application in industry, commerce and healthcare. We foster multidisciplinary working internally and collaborate widely externally.* (Imperial College London, 2013)

The translation of technology is also key to Imperial College's Strategy 2010–2014:

> *We will ensure that translating both into, and from, practice continues to remain an integral part of how we maximise value for society from our education and research.* (Imperial College London, 2013)

The mission statement, strategy and the main goals are supported throughout the company. The president, the board and directors of the research department all support the TTO activities. This is reflected in the organizational structures, reward systems and the many entrepreneurial and technology transfer initiatives.

Imperial College together with Imperial Innovations encourages researchers to disclose their technologies without distracting them from their research and education goals. Through their "Rewards to Inventors" scheme, the college offers inventors financial incentives. A significant proportion of net revenue from commercialization is shared directly with inventors, their Faculty and the College. Where the technology goes into a spin-out company, the inventor may choose to receive shares in that company, and may also benefit from consultancy or directors' fees, as well as a share of future licensing income.

The College also tries to increase the numbers of invention disclosures by pointing to the non-financial rewards. For instance, to see the results of your research leading into business growth, employment and wider benefits to society can be extremely rewarding. Secondly, intellectual property is recognized as an important component of research impact and can assist with obtaining many types of funding because translation is a key part of the College's strategy. Having a good financial and non-financial reward system is one thing, but communicating this to often busy researchers is another thing. The financial and non-financial rewards are clearly mentioned on the website of the College and the website of Innovations and are presented during the regular information sessions (Imperial College London, 2013).

Besides stimulating reward systems, Imperial College London has installed many entities that support the technology transfer process. Imperial Innovations is the TTO but is not responsible for attracting research funding and does not do any consulting activities as it is often the case for most TTOs. Two other entities at Imperial College support this: Imperial Consultants and Corporate Partners. Imperial Consultants is the UK's leading academic consultancy provider, connecting external organizations to the knowledge and resources of Imperial College London. They deliver confidential and commercial services including technical advice, expert witnesses and testing, measurement and analysis for industry, commerce and government worldwide. They provide a comprehensive professional service for Imperial academics who act as consultants. By taking on all the financial, contractual and administrative tasks, the academics can concentrate on the customer's project (Imperial Consultants, 2013). The Corporate Partnerships team facilitates bespoke industry collaborations with Imperial, tailored to the needs of all those involved. Imperial's current partnerships with industry range from highly specified contract research to pre-competitive knowledge commons to flagship industry-funded centers (Imperial College Corporate Partnerships, 2013).

> *Their main function is to form large and lucrative partnerships, most of the times with large companies, in order to attract large pots of money. For instance to fund a IBM computing lab ...* (Imperial College Corporate Partnerships, 2013)

5.2.2. Regional level: Financial community

An important source of funding especially for biotech and medical companies in London and throughout the UK are charity organizations. Some sources speak of more than 900 charity organizations and foundations which are based in the UK. They account for 59 billion euros on assets. One of the largest foundations in the world is Wellcome Trust, a global charitable foundation located in Euston, London and dedicated to achieving improvements in human and animal health. They want to support the brightest minds in biomedical research and the medical humanities and also support the technology transfer at universities by investing for instance in proof-of-concept funds. Their financial endowment is estimated around 17 billion euros. These charity organizations behave more and more like real venture capitalist and sometimes even impose strict milestones (London Medicine, 2013).

As mentioned above, London is one of the biggest financial districts in the world and many large banks have their headquarters in London. Consequently, many venture capitals are active in the London region. The venture capital community in London is huge. There are almost every day/evening opportunities to meet and network with venture capitals or business angels. London is also host for one of Europe's leading business angel investment networks called "London Business Angels" (LBA). LBA has the longest track record in the business, operating since the early 1980s. Since 2000 LBA has helped over 200 companies successfully raising over 60.5 mil euros. Over the last three years nearly 40% of the companies selected to pitch to LBA's investors have secured funding, despite an uncertain economic climate. Entrepreneurs raise money between 121,000 euros and 1.21 million euros (London Business Angels, 2013).

Imperial Innovations makes use of this window of financial opportunities. The venture team is responsible for this and maintains a large network of potential investors.

> *Ventures team has a large network throughout the venture capital world and angel investors as well. They have regular contacts with them and have a clear idea of what they are looking for. So... at the moment someone comes with this particular invention then they have a clear idea of which investors are most suitable. That is of course developed through time.* (Member of ventures team, Imperial Innovations, March 2013)
> *We do not only attract external capital, we also make use of the experience and the network of these investors. They could even help us with the recruitment of management team members.* (Deputy Director, Imperial Innovations, June 2013)

The venture team does not only have a strong network of co-investors, it also has its own funds to invest in spin-outs. The ventures team of Innovations identifies and invests in business opportunities arising from Imperial College and also from Cambridge, Oxford and UCL. Investments are made from the balance sheet of Innovations. There is no separate "fund" in the sense that many venture capital funds operate. This gives Innovations the freedom to invest according to its own criteria and means it is less subject to the whims, prejudices and fashions of the "market".

> *Innovations is not constrained by the VC mentality where they have to raise and spend money at time x and realize a return by time y. Innovations can take the long view if they need to. We can invest for 2 years in a company but we can also choose to invest for 10 years.* (Director of Healthcare Investments, Imperial Innovations, May 2013)

When Innovations finds it necessary, they also syndicate deals and bring in co-investors. One of the members of the venture team explains:

> *We can divide them in 3 main categories: the financial investor ... or private venture capitalist, corporate VCs (Johnson & Johnson, GSK,*

Pfizer,...) or the corporate themselves. In more early stage, we also co-invest with business angels or business angel associations (LBA, Cambridge angels,...). Very recently we invested in a mobile payment company together with several individual investors who had tremendous experience in that field. In the very beginning, if there is no clear proposition yet or there is no company formed yet, we use our development funds supported by charity organisations and/or governments. (Member of ventures team, Imperial Innovations, March 2013)

As stated above, Imperial Innovations can also use different proof-of-concept funds to support their spin-offs. The University Challenge Seed Fund is one proof-of-concept fund that was often used in the 2005–2010 period, together with the Low Carbon Seed Fund (2.9 million euros). The fund was supported by the government (Higher Education Innovation Funding, HEIF). They divided their investments into 15 University Challenge Seed Funds that were donated to individual universities or consortia. Each recipient university of a University Challenge Seed Fund had to provide 25% of the total fund from its own resources.

Today these funds do not exist anymore and were replaced by new funds. Innovations has various proof of concept and commercialization funds available including those supported by HEIF and via deals with the Wellcome Trust (Charity organization), Johnson & Johnson and other companies. Innovations calls them development funds. Innovations currently manages a range of these development funds for the development of therapeutics, bioengineering, engineering and software projects. This early stage investment supports the transition of ideas into marketable products. Innovations development funding can be used for a range of activities that will support the transition of an idea into a commercially viable project. For example, the funding could be used to pay for the development of prototypes, engage consultants, pay for external work to refine the idea or pay for consumables necessary to develop the project. The funding is not designed to pay for researchers' salaries, but should be used with the aim of developing data or improving product designs such that the output will inform a commercial decision. Typically, the funding will be used after the researcher has received translational grant funding, or to develop an idea before seeking a larger translational grant. Imperial Innovations operates a discretionary fund that makes available up to £50,000 for projects in any research area. In addition, they manage a number of funds aimed at specific areas.

5.2.3. Regional level: Government

The Department of Business, Innovation and Skills (BIS) is responsible to foster economic growth. The department invests in skills and education to promote trade, boost innovation and help people to start and grow a business. The department has a clear agenda (Smith, 2012):

- Strengthening the sharing and dissemination of knowledge (Issue: collaboration/competition).
- Supporting a coherent and integrated knowledge infrastructure (Issue: science base/Information infrastructure/Catapult centers).
- Encouraging business investment in all forms of innovation (R&D tax credits, corporate governance issues).
- Improving the innovative capacity of the public sector (procurement/services/technology selection).

More specifically, the department of Business, Innovation and Skills fosters technology transfer via the Higher Education Innovation Fund which is a funding program designed to encourage knowledge transfer in universities and other higher education institutions in England. In 2007, the BIS department set up an executive non-departmental public body called the Technology Strategy Board in order to stimulate technology-enabled innovation in the areas which offer the greatest scope for boosting UK growth and productivity. They promote, support and invest in technology research, development and commercialization. They spread knowledge, bringing people together to solve problems or make new advances. They also advise the government on how to remove barriers to innovation and accelerate the exploitation of new technologies. Below we explain the three main initiatives of the technology strategy board (Knowledge transfer networks, Biomedical Catalyst, catapult centers) (Smith, 2012).

Knowledge Transfer Networks (KTNs) are one of the Technology Strategy Board's key tools for enabling UK businesses to compete successfully at the forefront of global technology and innovation. They facilitate the UK's innovation communities to connect, collaborate and find out about new opportunities in key research and technology sectors. As a single overarching national network in a specific field of technology or business application, a KTN brings people together to stimulate innovation — from businesses of any size, research organizations, universities and technology organizations, to government, finance and policy. There are 15 KTNs and all 15 KTNs collaborate to form a "network of networks". Knowledge Transfer Partnerships aim to deliver the following benefits (Technology Strategy Board, 2013):

- Improving business performance through innovation and new collaborations by driving the flow of people, knowledge and experience between business and the science-base, between businesses and across sectors.
- Driving knowledge transfer between the supply and demand sides of technology-enabled markets through a high quality, easy to use service.
- Facilitating innovation and knowledge transfer by providing UK businesses with the opportunity to meet and network with individuals and organizations, in the UK and internationally.
- Providing a forum for a coherent business voice to inform government of its technology needs and about issues, such as regulation, which are enhancing or inhibiting innovation in the UK.

One of these 15 networks is the bioscience knowledge transfer network.

Another initiative is the biomedical catalyst — working with the Medical Research Council, they provide responsive and effective support to SMEs and academics looking to develop innovative solutions to healthcare challenges either individually or in collaboration. The Biomedical Catalyst is an integrated translational funding program (218 million euros) which supports UK businesses (SMEs) and academics looking to develop innovative solutions to healthcare challenges either individually or in collaboration. Three categories of grant are available: feasibility/confidence in concept award, early stage award and the late stage award (Technology Strategy Board, 2013).

The third initiative that is discussed here are the catapult centers. The government made an investment of over 242 mil euros in a network of elite Catapult Centres to help commercialize the outputs of Britain's world-class research base, bridging the gap between universities and businesses. This is a new network of physical centers designed to advance innovation in specific fields. The seven areas are: high value manufacturing, cell therapy, offshore renewable energy, satellite applications, connected digital economy, future cities and transport systems. Catapults enable business to access the best research and technical expertise, infrastructure and equipment to accelerate commercialization. Each center focuses on a field of technology or technology application in which the UK has particular academic and business strength. These centers will be an important part of the UK's innovation system, making a major long-term contribution to UK economic growth (Technology Strategy Board, 2013).

The technology strategy board considers the bioscience sector as one of the main focus areas. In this sector they want to focus on the characterization and discovery tools (commercial application of sequencing technologies focusing on genomic, phenotyping technologies, integration of -omics technologies, Development of biological imaging systems, biosensors, probes/markers, diagnostic platforms), production and processing (metabolic engineering, novel manufacturing processes for producing biological products and novel biological production systems, formulation and delivery approaches for biological products including biopharmaceuticals and functional foods) and bioinformatics (approaches to organizing, filtering and interpreting biological data including biological system modeling, data visualization, and user centered design) (Technology Strategy Board, 2013).

Besides the HEIF fund and the technology strategy board, the BIS department also supports a network of seven research councils.

One of the seven is the biotechnology and biological sciences research council (BBSRC). BBSRC has a unique and central place in supporting the UK's world-leading position in bioscience. BBSRC are an investor in research and training, with the aim of furthering scientific knowledge, to promote economic growth, wealth and job creation and to improve quality of life in the UK and beyond. BBSRC provide strategic funding to eight institutes. These institutes deliver innovative, world-class bioscience research and training, leading to wealth and job creation, generating high returns for the UK economy (Smith, 2012).

5.3. Implementing best practices in your region

Building a strong, effective and successful technology transfer system depends on internal and external factors. Best practices should also be implemented with great care and with a good knowledge of the universities' structure and ecosystem.

However, during the interviews with TTOs, a few guidelines were suggested in order to develop a strong technology transfer system. At the university level, you need to develop and stimulate an entrepreneurial culture. This could be by the entrepreneurial initiatives as explained above. In essence, you need to build an institution that respects and rewards entrepreneurial initiatives. This should be reflected in the reward system, the mission, organization structure and so on.

At the regional level, the TTO should build up a good relationship with investors and create an interesting ecosystem. By installing a team (cfr. ventures team at Innovations) with extensive previous investment experience at different types of investors, TTO should be able to build a strong network of investors and identify the right investors when different inventions and business opportunities arise at the university. The following are quotes from different team members at Innovations explain how success stories play a crucial role in the development of interesting ecosystems.

Look at how Silicon Valley started. They benefit from the expertise at Stanford University which was a huge resource. Shortly after a few success cases, investors started to group around Silicon Valley. Cambridge has also built such an ecosystem. Imperial has one. Oxford and UCL are catching up. Of course, everything starts with a technological solid university and a few success cases. (Deputy Director, Imperial Innovations, June 2013)

These success cases throw out new entrepreneurs. These success companies have many employees with equity options and when they realize an exit, these employees will be experienced enough and wealthy enough to become entrepreneurs. So that small tree throws up its seeds and a forest gets created in a certain period of time. These companies will also hire external experienced people and this is how the ecosystem starts to spread out. (Member of ventures team, Imperial Innovations, March 2013)

Imperial also had his success stories with Respivert and Thiakis and this helped to reach out to new investors saying we had success with a small amount of money, if you could provide us with more capital we could ... and so on... and this is also one of the reasons why we were able to raise another £140M. (Director of Healthcare Investments, Imperial Innovations, May 2013)

Imperial innovations also developed the ability to attract financial capital. One of the TTO offers a possible explanation for the successful investment rounds:

Innovations approached the investment community with a clear, interesting and compelling story. Investors, they want to spread the risk across a variety of different areas but if you invest in Innovations, you invest in the high-tech companies of the future coming from the four top universities from the UK. An investment in Innovations is an investment in the outputs of these four top universities. (Deputy Director, Imperial Innovations, June 2013)

The financial environment in London also made life a bit easier. As one of the members of the venture team concludes:

The development of Innovations is also due to the presence of strong financial community. It is one of the legs. Successful technology transfer is really influenced by the strength of the university, the entrepreneurs/human talent and the financial community/public market. (Member of ventures team, Imperial Innovations, March 2013)

Finally, the TTO evolution is also strongly influenced by the government. Imperial innovations could only be founded because the government had put the right legal framework in place that organizes the ownership rights between universities and stakeholders. There needs to be a robust set of IP Laws. The government should also carry out a policy which stimulates research and technology transfer. The HEIF and UCSF were all initiatives which stimulated the growth of Imperial Innovations and many other TTOs around the country.

5.4. Future improvements

- Although Imperial Innovations invests in spin-outs from its own balance sheet, Imperial College wants to further invest in different types of proof of concept and development funds.

5.5. Best practices

- World-class research institute which delivers high potential technologies.
- An entrepreneurial university that respects and rewards entrepreneurial initiatives.
- A ventures team with investment background that builds a network of investors and has a clear compelling story to convince investors.
- Development funds to support the transition of ideas into marketable products.
- Government initiatives to stimulate technology transfer: Technology strategy board and Higher education innovation fund.

5.6. References

Papers, reports, books, conference presentations and websites

City of London (2013). Retrieved from: www.cityoflondon.gov.uk. Accessed on 04.02.2013.

Imperial Innovations (2013). Retrieved from: http://www.imperialinnovations.co.uk. Accessed on 12.01.2013.

Imperial College London (2013). Retrieved from: www3.imperial.ac.uk. Accessed on 13.01.2013.
Imperial College Corporate partnerships (2013). Retrieved from: http://www3.imperial.ac.uk/corporatepartnerships. Accessed on 13.04.2013.
Imperial Consultants (2013). Retrieved from: http://www.imperial-consultants.co.uk. Accessed on 13.04.2013.
London Medicine (2013). Retrieved from: www.londonmedicine.ac.uk. Accessed on 13.01.2013.
London Business Angels (2013). Retrieved from: www.lbangels.co.uk. Accessed on 16.04.2013.
Penfold, H. (2012). *How Innovations Started and Grew to What it is Now.* ETTbio site visit, Imperial College London.
Smith, K. (2012). *How Government and Technology Strategy Board Supports Technology Transfer in the UK.* ETTbio site visit Imperial College London.
Technology Strategy Board (2013). Retrieved from: www.innovateuk.org. Accessed on 08.02.2013.

Case study interviews

- 28.03.2013, Member of venture team, Imperial Innovations, London.
- 13.05.2013, Director of Healthcare Investments, Imperial Innovations, London.
- 21.06.2013, Deputy Director, Technology transfer team, Imperial Innovations, London.

6

CASE STUDY 6: The Industrial Research Fund

Tom Guldemont, Thomas Crispeels and Ilse Scheerlinck
(Vrije Universiteit Brussel, Vesalius College, Belgium)

6.1. Setting the scene

The Industrial Research Fund (Industrieel OnderzoeksFonds, IOF) is one of the two programs to support strategic basic research in Flanders. It aims at the economic exploitation of knowledge production at universities, by building up applied science portfolios at universities, and stimulating university–industry linkages. The rationale of IOF is to address the gap between research and the economic exploitation of the knowledge produced at universities. IOF aims to gear universities towards more application-oriented research. The strong position of universities in the Flemish innovation system has made it possible for them to foster the launch of this government program for intra-university research whereas inter-university research is supported through the "Strategic Basic Research financing channel" (SBO).

The IOF program was established by the Flemish government in 2004 as a means to create a more permanent framework for the support of strategic basic research and valorization-oriented projects at the universities of the Flemish community. Trying to meet the overall academic demand for a more attractive and flexible research environment and more diverse types of researchers, this fund enables the development of a more devised long-term policy for strategic and applied research programs at universities.

6.1.1. Flanders

The region of Flanders is one of the three regions in Belgium. Situated to the north of Brussels, the capital city of both Belgium and Flanders, it is home to Belgium's Dutch-speaking community. Its approximately 6.3 million inhabitants live at the crossroads of the Netherlands, Germany, France and the United Kingdom, on a surface of $13,500\,km^2$ (2/3 of Silicon Valley). In addition, its capital Brussels is home to the European decision-making centers. Besides tax incentives, these are two crucial factors for a company when making an investment decision and choosing a location in Europe.

Flanders is a successful life sciences and biotechnology region with many biotech companies located within a small geographical area. Local and international investors are providing the capital for their continued growth. The full range of competences required to bring new and innovative products to the market are locally available. In Flanders, a life sciences investor will find a unique interplay between business, universities, research centers and hospitals. Good interaction and extensive networking between the various public and private life sciences players provides a dynamic environment, rich in innovation and knowledge sharing, in which new companies are constantly being added to a fast-growing life sciences cluster.

Belgium has four large, export-oriented clusters: chemicals, biopharmaceuticals, plastics and jewelry. Belgium is the world's second largest exporter of biopharmaceutical products. Strong factor conditions include a dense network of 167 hospitals and high-quality educational institutions (as measured by citations). This supports one of the highest pharmaceutical R&D re-investment rates in Europe. Belgium's favorable business context includes the fastest approval process in Europe for clinical trials (Phase I in less than two weeks). As a result, Belgium is the world's number one location for clinical trials per capita. The cluster also benefits from strong supporting industries such as biotech, chemicals and logistics, which complement industry activities in R&D, manufacturing and distribution (Flandersbio, s.d.).

Flanders houses the biggest R&D hub for plant biotech — headed by the Flanders Institute for Biotechnology (VIB) — and emerging industrial biotech, with one of the largest integrated bio-energy production complexes in Europe — Bio Base Europe.

6.1.2. The Flemish Department of Economy, Science and Innovation (EWI)

The EWI was established in 2006 after the Flemish governmental reorganization within the scope of Better Administrative Policy. The EWI Department prepares, monitors and evaluates policy in the economy, science and innovation policy area. In doing so, the main aim is to develop Flanders into one of the most advanced and prosperous regions in the world. Its driving forces are the promotion of excellence in scientific research, an attractive and sustainable business climate and of a creative, innovative and entrepreneurial society.

The interplay between the economy, science and innovation holds out unique opportunities to develop a future-oriented long-term strategy. Working in a climate of openness and cooperation, EWI tailors its ideas to those of their partners in its own policy area and with the other policy areas. The main tasks of EWI are to formulate policymaking proposals both proactively and at the request of the competent ministers. EWI also monitors the policy cycle by critically evaluating the effectiveness of policies that have been implemented and adjusting the approach on the basis of objective results.

EWI is connected to relevant stakeholders and governmental agencies in Flanders such as Enterprise Flanders (Agentschap Ondernemen), The Research Foundation — Flanders (Fonds Wetenschappelijk Onderzoek, FWO) and Agentschap voor Innovatie door Wetenschap en Technologie (IWT) (EWI, 2013).

6.1.3. IWT

IWT is the government agency for Innovation by Science and Technology, and was established by the Decree of 23 January 1991 of the Flemish Government. The agency helps Flemish companies and research centers in realizing their research and development projects. IWT offers them financial funding, advice and a network of potential partners in Flanders and abroad. It also supports the Flemish government in its innovation policy (IWT, 2013).

6.1.4. *Vrije Universiteit Brussel*

The Vrije Universiteit Brussel (VUB) is the coordinating university of the University Association Brussels (UAB), which consists of a strategic partnership between the VUB, the Brussels University Hospital (UZ Brussel) and the Erasmus University College (Erasmushogeschool Brussel, EhB) as key partners.

Vrije Universiteit Brussel is one of the five Flemish universities, employs approximately 2,900 people, including approximately 1,500 FTE R&D staff, and has around 11,000 students. VUB consists of eight faculties, of which two are related to life sciences and biotechnology: the faculty of Science and Bio-engineering sciences and the faculty of Medicine and Pharmacy. Furthermore, the Vrije Universiteit Brussel is connected to the Flanders Interuniversity Institute of Biotechnology (VIB).

The university has two campuses. The main campus in Etterbeek houses seven faculties, while the medical campus in Jette is home to the faculty of Medicine and Pharmacy and to the UZ Brussel, which is the university hospital. The UZ has 721 beds, 28,200 hospitalized patients and 3,400 employees (VUB, 2012).

6.1.5 *Interactions between universities and industry*

The Flemish government made the development of the knowledge economy as its priority and positioned the knowledge institutions as an engine of innovation. Universities and their association partners are encouraged by the government to cooperate with industry, with the goal of boosting innovative knowledge-building and technology transfer. Within the Flemish universities, depending on the university, between 15 and 30% of all R&D expenditures are funded by the private sector. In 2009, the privately funded part of the HERD (Higher Education R&D) amounted 16.1% for Flanders (ECOOM, 2011, p. 99), a percentage with which Flanders achieves an absolute top score in OECD-context with regard to the involvement of industry in research funding of higher education. Also in the field of exploitation of applied research through industrial/economic and social projects, patents and spin-off companies, the academic sector experienced a growth trajectory through recent years. The IOF was an important incentive in this evolution. In this way, the government satisfies the need of completing the necessary funding gap between obtaining innovative research on the one hand and its marketing on the other.

6.2. Identifying best practices

The Industrial Research Fund (Industrieel Onderzoeks Fonds, IOF) is a structural funding from the Flemish government that provides universities the ability to perform an autonomous policy with a view to building a portfolio of application-oriented research and promoting the interaction of the associations with industry, and all this based on their own strengths and opportunities. The IOF consists of a sealed envelope, to be spread over five universities: University of Leuven, Ghent University, Free University of Brussels, University of Antwerp and Hasselt University.

Within Flanders there are two programs supporting strategic basic research: Strategic Basic Research (Strategisch Basis Onderzoek, SBO) and the IOF. Whereas SBO is aiming at cooperative research between universities and research institutes, with subsidies being awarded by IWT after an open competition (with external expert judgment), the IOF is divided over the Flemish universities every year, based on distribution formula with several criteria (number of personnel, number PhD theses, publication output, number of patents, participation of university in IWT projects and FP projects, number of spin-offs). Every university then has an intra-university competition to award the IOF-funding to projects.

From the beginning the IOF has contributed to the vision on the implementation of the business-oriented valorization policies of each Flemish university. Today that valorization policy is identified and even characterized by an own specific IOF operation, which is the central engine in building a portfolio of application-oriented knowledge for economic purposes.

Notwithstanding the fact that each university has its own policy and places its own emphases, the IOF harmoniously matches in all institutions with the valorization policy and interface operation. The difference between universities in spending and implementation of the IOF funds is partly due to the differences in the IOF budget that the different associations receive.

6.3. The history and evolution of the IOF is closely linked to the Interface Offices at the Flemish universities

By decree of 19 December 1998, the Flemish government started to provide subsidies to universities in order to establish Interface Offices at these institutions. These offices are charged by the Flemish government to implement interface activities in order to promote:

- Cooperation between Flemish universities and businesses
- The economic valorization of university research
- The creation of spin-off companies by universities.

On 28 May 2004, the Flemish government approved the establishment of the IOF. IOF has been based on ad hoc regulation in 2004 and 2005 but has changed in a formal program (Decree) in 2006, learning from the experience in previous years. In this formal program, the criteria which determine the distribution of IOF funds over the universities have been gradually changed.

The Decree of 5 June 2009 of the Flemish government joined the Interface Offices and the IOF, started the current cycle (2009–2014), providing the new objectives and framework for the management of the offices and the IOF. However, the interface activities and the IOF each have a separate funding base and allocation key.

The Interface subsidy, for a total of 2.8 million euros (2012), is currently allocated to the associations according to the Interface allocation key, taking into account the scientific personnel for the period 2003–2007.

6.3.1. The workings of the Industrial Research Fund

The rationale of IOF is to address the "European Disease", i.e. the gap between the (often excellent) research and the valorization of the knowledge produced at universities. One way to deal with this gap between research and industry is to stimulate application-oriented research at universities. IOF must gear universities towards more industrial relevance. It does not target specific research and technology fields.

For inter-university research the SBO program was set up. The strong position of universities in the Flemish innovation system made it possible for them to also create a government program for intra-university research. In this program the universities themselves decide where to spend the money on. The distribution of the funds over the universities is however output-related, and based on valorization indicators (like patents, number of spin-off companies, etc.).

The objectives of the IOF can be summarized as follows:

In the short to medium term, the IOF

- should stimulate the interaction between the university association and the industry
- should build up a portfolio of applied knowledge at the association.

In the middle to long term, the IOF should result in

- a better alignment of strategic research and applied scientific research with economical needs
- the application and valorization of the built up knowledge portfolio in the industry.

6.3.2. Allocation key

Every Flemish university manages its own Industrial Research Fund. The university receives funding according to its output performances, and then sets up an intra-university competition to award the IOF-funding to projects.

A set of seven weight input and output parameters were developed to create an "allocation key" that determines the share of IOF-subsidy for each association. The seven parameters include:

- PhD degrees
- publications and citations
- industrial contract incomes, such as IWT-projects
- contract revenues from the Framework Program of the European Union
- patents
- spin-off companies
- the number of scientific personnel.

In 2009, the Decree of the Flemish Government concerning the support for industrial research funds and interface activities of the associations in the Flemish Community, entered into force. Since then, a second phase started, in which a greater emphasis was placed on the achievements (output) that can be directly linked to valorization: industrial contract revenue, patents and spin-off companies.

The other parameters (e.g. scientific workforce, total allocated sum of IWT funds) disappeared gradually from the allocation key or became less important (e.g. doctoral degrees, publications and citations). The science-oriented parameters were included in the first years to give the opportunity to all universities to build up a strong Interface Office. Indeed, some of them already had a well-developed Technology Transfer Office (TTO) long before the Flemish government started to provide subsidies for this. Others only started at the end of the nineties and did not have good scores yet on valorization-oriented parameters. In order to not discourage these institutions, the science-oriented parameters had their place at the early stage.

Each association manages the IOF Funds. The exact management structures differ, however, the IOFs need to have the following structure by decree. Each association has an IOF Council that advises the association board on the allocation of the budget of the IOF. The IOF Council consists of at least 12 members of equally represented groups of the partners of the association (universities and colleges) and of companies. The academic or association board decides where the IOF budget is to be allocated. However, the IOF always needs to be allocated to strategic or applied research.

6.3.3. Review and follow-up

6.3.3.1. Strategic plan

Every five years the associations develop a strategic plan for the IOF, including the strategic and operational objectives they want to achieve with the IOF. The Flemish minister for Innovation can define additional rules for the strategic plan, particularly for cooperation initiatives between the associations themselves and in the context of the Flemish Innovation Network. The associations have to send their new strategic plans, in duplicate, to the Flemish Minister for Innovation, before September 1 of the year when the current strategic plan expires.

The Department of Economy, Science and Innovation (EWI) evaluates the plan on completeness and compliance within three months after the deadline. For the purposes of this review, the Institute for the Promotion of Innovation by Science and Technology in Flanders (IWT) assesses the plan within two months after the deadline on how one wants to come to an optimal cooperation between the associations themselves and in the context of Flemish innovation network. In cases where the conditions are not met, the associations can make adjustments in the submitted, but not yet approved, strategic plans in consultation with the EWI. On the advice of the EWI, the Flemish Minister for Innovation approves the plan within three months after the deadline.

6.3.3.2. *Information duty*

Annually and before April 30, the associations report to the Flemish Minister for Innovation about the achievements in the implementation of the strategic and operational IOF-objectives and the cooperation between the associations. This report is submitted to the EWI and the IWT. Within three months after the deadline and on the advice of the EWI, the Flemish Minister for Innovation assesses the report according to completeness and compliance with the governmental decree and the strategic plan.

6.3.3.3. *Monitoring and evaluation*

Every five years the Flemish government reviews the global IOF and interface activities with a view to possible adjustments. At least the following issues are examined:

- the application of the Governmental Decree
- the achievements of the strategic plans
- the scientific output and the results in terms of economic exploitation.

6.3.3.4. *Review of progress*

There is no direct follow-up on the progress, as the IOF is no traditional program with projects, it is merely basic funding for universities to conduct applied or strategic research. However, the output of this research determines to some extent the allocation of IOF funding in upcoming years, keeping in mind that the allocation is based on the parameters mentioned above, namely PhD degrees, publications and citations, industrial contract incomes, contract revenues from the Framework Programme of the European Union, patents and spin-off companies.

6.3.3.5. *The approach of VUB*

The IOF fund is an integral part of the university's valorization policy, which is made available via the VUB Industrial Research Fund. These resources are actually used by the IOF Council within the broader framework of efforts that should strengthen interactions between higher education institutions and economic actors. The appointment of own valorization managers in research groups who, in cooperation with the Technology Transfer Interface (TTI) business developers, take control of the valorization roadmaps of research groups, has led to a sharp increase in technology transfer activities.

The coaching and support of mandates and projects financed by the IOF is the responsibility of the TTI. The TTI also assists in the preparation and follow-up of research projects aiming at industrial research activities, projects for industrial development activities, and projects for strategic basic research of industrial relevance in collaboration with and funded by the IWT and IWOIB.

6.3.3.6. IOF Council

According to the decree of the Flemish Government, the IOF Council consists of at least 12 members and is divided into three sections:

- staff of the university partner to the association in issue
- staff of the colleges that partner in the association in question
- business representatives.

The three sections each consist of at least one-quarter of the members of the IOF Council.

The VUB IOF Council is responsible for the following tasks:

- The development of an own IOF-policy, aimed at developing a portfolio of potential application-oriented knowledge for economic purposes.
- Advise the university authorities regarding the best use of granted IOF-resources in accordance with the short and medium objectives as formulated in the current Decree.
- Develop the call and selection procedures for granting IOF funds.
- Develop a feedback procedure for applicants which were not allocated with IOF resources.

- Develop arrangements for conflicts of interest.
- Develop support and/or policy measures regarding the allocation of resources.

The mandate of the IOF Council members lasts one year. Renewals of mandates are possible. Changes in the composition of the council have to be submitted to the university authorities. The members of the IOF Council and the Secretariat are bound by strict confidentiality with regards to the data provided by the departments/research groups in the context of their IOF application. To that purpose, each member of the Council and the members of the TTI sign a confidentiality agreement.

The Secretariat of the IOF Council is provided by the VUB Technology Transfer Interface under whose responsibility the daily management of funded mandates and projects is being conducted.

Conflicts of interest may arise when members of the IOF Council (regardless of their status or employer) have the ability to influence practices or decisions of the IOF Council, so that they, the groups where they belong to or persons to whom they are connected, are able to obtain advantage out of it. Such benefit may have a financial or another impact, and can be acquired directly or indirectly (through a related person).

6.3.3.7. *VUB internal IOF regulation*

The "VUB IOF resources" are defined annually by the Flemish government, based on a performance-oriented allocation key. The reference period for these parameters is always a sliding time window from year n-6 to the year n-2, where n is the year in which the key is used. As such, the university annually hears from the EWI what amount it can include in the of the university budget. The utilization of these IOF funds is subject to a number of criteria set by the government. They are not recurring credits, but allocations based upon a project application after screening by the IOF Council and after approval by the University Board.

The IOF Council will allocate funds under the following conditions:

- Allocation by the University Board after an open call, based upon motivated advice of the IOF Council.
- At least 30% of the IOF resources should be spent on research mandates of indefinite duration.
- Up to 10% is intended to cover expenses (operating expenses and labor costs) related to the management and operation of IOF. With these resources, the costs of the screening by external parties are borne and people are appointed within the TTI, on the one hand for the management of the IOF and the other hand for business development, to support all researchers.

The other IOF funds can be used for:

- Operating expenses, equipment costs and labor costs for research projects, with a minimum duration of one year
- Project expenses to support IOF mandates (VUB internal documents).

6.3.3.8. 2004–2011: Program funding

From 2004 to 2011, the IOF funds at the VUB were allocated as program funding and assigned to so-called IOF-cores and growers, research units with potential for growth towards excellence. The budgets of 220,000 and 110,000 euros per year respectively, were granted for a period of five years to these research groups or departments.

In case of an IOF-core, the lab typically exceeds the average scale of a standard university's research unit. These groups are able to conceive a detailed long-term roadmap and vision and have strongly motivated how the extra IOF funding might contribute to their proprietary valorization strategy and would make an effective difference in terms of knowledge and technology transfer. Furthermore, they are carrying out a substantial amount of outstanding strategic research that leads to new application-oriented inventions with economic potential, enabling the establishment of a portfolio of potentially applicable and transferable know-how with economic and societal value.

The assessment of the submitted applications for program funding was primarily based on the actual output of the applicants on the IOF parameters and the proposed growth trajectory as a function of the expected valorization outcome. In total, six Cores and three Growers were funded as IOF program at the VUB.

In 2011, the IOF policy at the VUB was changed. However, the already granted IOF program financing for Cores and Growers was allowed to keep running for five years, as had been already set out. This was subject to annual positive evaluation by the IOF Council (Internal documents VUB).

6.3.3.9. 2011–...: Project funding

Since 2011, new applications can be subsidized as Group of Expertise in Applied Research (GEAR). Also as of 2011, the IOF means are allocated to projects instead of to programs. The IOF Council allocates funds as project financing (as opposed to program financing) to either research groups, departments or consortia that:

- transcend the average scale of the university in the field of strategic and applied research, and
- that can submit a detailed valorization path where the grant of one (or more) IOF-mandates can make an effective difference to realize the roadmap that is put forward.

These groups have to prove their existing track-record in valorization activities, as can be deduced from their performance on the IOF parameters "industrial contract & licensing revenues, patents and spin-offs". These research groups, departments or consortia receive project funding for three-years.

Each year the GEAR groups submit a brief report to the IOF Council that describes the evolution of the IOF parameters. This report is orally defended by a spokesperson of that GEAR (Vrije Unversiteit Brussel, 2012).

6.3.3.10. Proof-of-concept funding

Since 2009, and in addition to the longer-term funding cited before, the Industrial Research Fund at VUB also funds proof-of-concept projects, e.g. research projects that are in the transition phase from proof of principle to proof of concept. Each year, the IOF Council reserves an amount of at least 100,000 euros for this proof-of-concept financing. This requires a separate call to be launched. The project proposals are screened by the TTI and a funding recommendation is submitted to the IOF Council, which submits at its turn a proposed decision to the university Board. The management of the resources falls under the control of TTI, which monitors budgets and spending ex-post and which may recover any unjustified spending.

6.3.3.11. Reporting

The IOF Council reports annually about his work to the Board of Directors of the Vrije Universiteit Brussel (in the context of the global VUB report regarding research activities). This report provides an overview of the resources used and the results in the field of industrial research funded by the IOF resources.

The University Board reports annually to the Flemish Minister for Innovation on the use of the IOF funds, analogous to the BOF spending. Costs and revenues are shown in the university's accounting and financial statements. Any IOF funds that are not assigned at the end of the calendar year can be transferred to the budget of the following year and still be used for the same purpose. The president of the IOF Council submits an annual report to the Flemish Minister for Innovation. This report is prepared by the TTI and is part of a larger report concerning the achievements of the TTI.

6.3.3.12. Impact of IOF at VUB

As from the beginning, the VUB has chosen not to distribute the resources over as many researchers as possible, but to invest in research groups that already achieved a substantial output as measured by the IOF parameters, or in groups which have that potential and could submit a long-term roadmap. These groups can then make the choice to embed existing postdoctoral researchers or to recruit valorization managers who bring new expertise to the group. The IOF Council expects that the greatest added value with the IOF funding can be achieved with this approach. Underlying reasons are also related to the growth or maintenance of the share of the VUB in this sealed envelope funding.

The evaluation has shown that the granting of a program funding to these groups indeed has led to an increased performance in all cases. Depending on the valorization roadmap, which the group itself chose to deploy, very different results were obtained:

- In the period 2005–2009, IOF Core "Electronics and Informatics", realized three new spin-off companies.
- IOF Grower "Robotics and Applied Mechanics" recruited a business developer with many years of business experience, who introduced priorities in the many valorization tracks that were running in the research group.
- IOF Grower "Toxicology and Dermato Cosmetology" decided to bring in more industrial partnerships and oriented a couple of their research projects in such a way that a freedom-to-operate could be kept in for them strategic research lines.
- IOF Core B-Phot (Applied Physics and Photonics) chose to focus its program funding for the first five years on anchoring strategic researchers, who in this way could devote more time to set up contracts with industry, which resulted in an impressive multiplication of revenues for the group.

Of course, the choice of the VUB to invest in the most successful groups is an explanation for these results. However, it is undeniable that with the IOF funding as additional injection, a significant increase in return could be realized. The requirement for a funding request (to submit a well-founded valorization roadmap which is tested by external experts) means that many groups — often for the first time — have built a long-term strategy.

6.4. Implementing best practices in your region

6.4.1. *Implications for TTO*

- The TTO is responsible for the workings of the IOF within the university. It is the link between the university and the Flemish government.
- TTO has to supply the Flemish government with the output score on the IOF parameters. Collect the appropriate data from all groups however, is often a very complex task. A good administration is a prerequisite for this.
- The TTO has to set up an IOF Council within the university structure.

6.4.2. *Implications for university/research organizations*

- Set up an Interface Office, in other words a TTO, with skilled people.
- Set up and organize an IOF Council, which has a long-term mindset.

6.4.3. *Implications for policy makers*

- Free up resources in the total budget of the government.
- Set up an official decree to establish IOF regulation (and Interfaces if necessary).
- Set up databases where some of the parameters can be smoothly monitored.
- Carefully consider the parameters used to allocate means, define the goal of this policy instrument.

6.5. Future opportunities

- Effect is only visible in the long term: The IOF was only established in 2004. Given that the development of science and technology takes some time, the positive impact of the IOF showed

above, and the fact that IOF funded groups have a better valorization track-record than other groups, is in the first instance an indication that the IOF funds are allocated to appropriate groups. This is especially true within universities, where projects are running for three or five years and it is sometimes difficult to identify a leverage effect.
- Limited contribution to total R&D funding: the VUB has a total R&D budget of approximately 70 million euros, where the IOF support represents only 1.5 to 2 million euros, which is not substantial.
- IOF parameters are limited: parameters are measurable, but often very limited, as they do not give the total picture:
  - There will always be a discussion on which (additional) parameters should be used and which are adequate.
  - Having a low score on the IOF parameters does not necessarily mean that there is no valuable valorization work being done at that university.
  - The parameters are used to calculate the allocation key, but also to measure the impact on economy and society. However, they don't guarantee an increase in employment. For example, with a view on its IOF parameter output, a university could prefer to establish ten service spin-off companies each employing one person, than to create a spin-off that immediately employs 25 people.
  - The IOF parameters only take the absolute output of universities into account. They don't look at the efficiency and the size of universities. A university with 2,000 researchers that creates ten spin-offs will receive more IOF funds than a university with 1,000 researchers that starts up nine companies, even though it is less efficient. In this context, the current system does not reward the best groups in terms of valorization. This could be improved by looking at relative parameters, where output is divided by the number of researchers.
- Insecurity about the yearly IOF support:
  - An improved score on the IOF parameters does not guarantee a larger share of the IOF subsidy. If other universities have

realized better improvements, the received IOF funding can even decrease.
- o The total IOF subsidy is dependent on the government budget. Even if all universities were successful in the previous years, it is possible that the government cannot increase the total IOF support.

6.6. Best practices

- The IOF is unique: it is the only internal financing possibility at the disposal of universities that can only be used for valorization. If there was only money available for fundamental research, research groups would probably only do fundamental research.
- Universities can use the IOF money in relative freedom (but within the boundaries set by the Flemish Government). They can freely choose their strengths and policies, as long as the objectives are reached in the end. To avoid a fragmentation however, funded projects should have a budget of minimum 50,000 euros per year and a lifetime of at least one year.
- IOF is closing the funding gap. It bridges the gap between proof of principle and proof of concept, often the first milestone that is required by (risk averse) seed capital funds or venture capitalists.
- The sliding window of five years in the calculation of the parameters is a good approach to correct for outliers and excesses.
- The parameters have their limits, but the existing ones are certainly not unfounded, they are oriented at valorization.
- IOF increases the awareness of valorization and promotes a change in the attitude on valorization orientation of researchers. Making researchers work towards technology transfer output parameters, it has a sensitizing, educational effect on them. One IOF mandate holder can have a profound effect on the culture within a research group. This person can attract extra funding aimed at valorization of research results.
- IOF contributes to the realized valorization output of universities.
- Good results of the IOF are a good argument towards politicians to free up more resources for the IOF budget.

6.7. References

Papers, reports, books, conference presentations and websites

Besluit van de Vlaamse Regering betreffende de ondersteuning van de Industriële Onderzoeksfondsen en de interfaceactiviteiten van de associaties in de Vlaamse Gemeenschap 2009 (Flanders) B.S.23/07/2009 (Belgium).

ECOOM (2011). Vlaams Indicatorenboek WTI. Retrieved from: https://www.ecoom.be/sites/ecoom.be/files/downloads/indicatorenboek2011.pdf. Accessed on 03.04.2013.

Flandersbio (s.d.). 10 Reasons to set up a Life Sciences R&D entity in Flanders, Belgium. Retrieved from: http://flandersbio.be/files/Factsheet_FINAAL.pdf. Accessed on 03.03.2013.

TTO Flanders (2011). Het industrieel Onderzoeksfonds: Een balans 2006-2010, Hefboom naar technologietransfer en samenwerking met de industrie. Retrieved from: http://homes.esat.kuleuven.be/~bdmdotbe/newer/documents/IOF_BOOK_def_v8-1.pdf. Accessed on: 05.03.2013.

VUB (2011). Internal IOF reglementation.

Vrije Universiteit Brussel (2012). About the University and Facts and figures. Retrieved from: http://www.vub.ac.be/english/home/about.html. Accessed on: 10.03.2013.

7

CASE STUDY 7: Regional Innovation Vouchers as an Effective Tool for Supporting Technology Transfer

Regional Development Agency Ostrava
(RDAO, Czech Republic)

7.1. Setting the scene

According to the Regional Development Agency Ostrava (RDAO, 2012), the Moravian-Silesian Region has been chiefly renowned for its importance to heavy industry — mechanical engineering, metallurgy and mining — which still have a substantial influence on the region's character today. Naturally, therefore, the majority of the research and development capacities are associated with big companies in these industries. However, there are other promising domains such as IT, electronics, biotechnology and environmental technologies, where research and development activities are concentrated primarily at the VSB — Technical University of Ostrava, University of Ostrava and University Hospital Ostrava as well as several companies involved in the development of sophisticated products. As for public research and development, the VŠB (Technical University Ostrava) plays a unique role: it is currently implementing several projects of new research centers, supported from the Operational Programme Research and Development for Innovation, worth a total of 4 billion CZK.

Besides the aforementioned universities, the Moravian-Silesian Region is the seat of other public universities and colleges (the Silesian University in Opava, College of Social and Administrative Affairs in Havířov, and Business School Ostrava) attended together by approximately 40,000 students every year. Most of the research sites and colleges were established in response to the economic development caused by the changes and restructuring in the latter half of the 20th century. Very interesting is also the fact that the most important industrial sectors in the Moravian-Silesian Region are already organized in clusters, which gives the Region a new profile and makes it much simpler for investors to access individual subcontractors. In this respect, this is the leading Region in the Czech Republic. In recent times there are ten functional clusters in the Moravian-Silesian Region (RDAO, 2012; Moravian-Silesian Region, 2014).

One of the key problems facing not only the Moravian-Silesian Region but all regions is how to promote cooperation between universities or research institutes and commercial companies. Such cooperation is essential if the results of research and development are to be transformed into effective, practical commercial solutions. The Regional Development Agency is helping to address the barriers standing in the way of academic/commercial cooperation as part of its tasks to implement the Moravian-Silesian Regional Innovation Strategy 2010–2020.

If the Region is to be competitive in the future, it needs to support the development of innovative companies. Harnessing the potential of innovative companies, universities and research institutes will bring a competitive advantage. The science and research subsidies are an effective way of supporting the start-up of projects in the research and development sector. It offers mostly SMEs a way of setting up cooperation with universities and research institutes, which may well lead to major long-term partnerships in their particular field (RDAO, 2012; Moravian-Silesian Region, 2014).

Therefore, a subsidy program supporting projects involving collaboration between manufacturers and universities or research institutes in the Region was created — the innovation voucher.

7.2. Identifying good practices

Within the survey we conducted during the preparation of the Regional Innovation Strategy of the Moravian-Silesian Region, we found out that the universities, research institutes and companies are very interested in mutual cooperation and declared that the main motivation factor for them is the expertise level increase and financial benefits of getting new contacts. However, most respondents indicated that the Region's universities are actually rather passive when it comes to seeking collaboration with the industry, often only responding to initiatives or questions coming from industrial partners.

In terms of funding for research, development and innovation activities at universities and research institutes, it all heavily depends on the public financial resources. According to respondents, the system of public funding of research and development activities is administratively complex and biased towards the more developed regions in the country. Additionally, the European Structural Funds are among the main financial sources of financing the R&D activities currently. The representatives of universities negatively evaluated the high administrative cost related to the preparation and implementation of the EU projects in comparison with national programs. Only a small portion (around 10%) of funding for research and development activities of universities and research institutes comes from private sources.

All the aforementioned issues were the reason to include the support of collaboration between companies and research institutes into the implementation of the Regional Innovation Strategy scheme. The main objective was to prepare a financial tool with minimal administrative burden for developing mutual trust and cooperation between regional companies and research institutions. Companies thus have the opportunity to test collaboration with selected research teams in quite easy administrative procedures.

7.2.1. *Specific description*

Innovation voucher is a financial instrument supporting cooperation between companies and institutions with research capacities. It is an integral part of the Regional Innovation Strategy of Moravian-Silesian

Region. The aims are to help break down mutual barriers and prejudices between companies and research institutions and develop new collaborations in the future. On one hand, competitiveness of companies might be strengthened; on the other, the commercialization of research results of the research institutions might become more effective. Innovation vouchers are a subsidy provided to a company in order to purchase a research service from a research institute. This service is based on knowledge transfer, i.e. transfer of knowledge of a scientific or technological nature that is new for the company and is not commonly available. Purchased knowledge must, at the same time, lead to the strengthening of companies' competitiveness, mainly through innovating its product, process or service (Stejskal & Matatkova, 2012).

Below are the basic facts regarding the innovation voucher:

- Financial sponsor — financial sponsor of the subsidy program is the Regional Authority of the Moravian — Silesian Region.
- Implementing body — implementing body is also the Regional Authority, namely the Department of Regional Development. However, there are also other options how to implement this kind of program. It is quite often that there is an implementing agency which is responsible for design of the program, implementation and coordination — e.g. collecting of incoming application, selecting of voucher recipients, paying out subsidies, supervising and evaluation of the program results.
- Research institutes (knowledge providers) — knowledge providers are institutions from all around the world that have research capacities that have signed a contract about participation on the program. Each knowledge provider appoints at least one contact person who coordinates the project at his/her institution.
- Voucher recipients[2] — an innovation voucher can be awarded to a legal person established for a business purposes with the address in the M-S Region.

---

[2] The innovation vouchers may be opened to companies from all around the world or vice versa to research institutions from other regions. It depends on the chosen model.

7.2.2. Evolution of the innovation voucher

Differences between academic and commercial spheres are, were and will be. The issue of differing perceptions of reality by universities (or more precisely, academics, scientists and researchers) and by firms has already been paid a lot of attention as well as breaking barriers, building bridges between the academic and commercial world, etc. Nowadays, in the context of the ongoing reform of higher education in the Czech Republic, there is continuously increasing importance of the so-called third role of universities. Also the realization of vast projects under the Operational Programme Research and Development for Innovation and set rules of sustainability of cooperation between companies and universities are still very current. The enterprises are still distrustful of cooperation with the academic sector due to lack of motivation regarding the implementation of "commercial projects". They've met with phrases from academics such as "scholars must teach". Fortunately, this paradigm has changed in recent years and universities are getting more active in terms of market-oriented collaboration.

A sense of innovation vouchers is precisely to overcome these "barriers" and thus eliminate the initial distrust of this type of cooperation. Enterprises can try to cooperate with knowledge providers on smaller projects and change the ingrained distrust. Another benefit is a possibility of reimbursement of the costs incurred. The idea was that good experience develops further cooperation. Finally, it is also an opportunity for knowledge providers to develop their business-oriented internal mechanisms.

Nascence of innovation voucher in the M-S Region has been ascribed to the former Deputy President of the Region responsible for the regional development who is also academic with a background in economic development. It is very important to have the support of an enlightened policy maker who is able to push this kind of initiative further and negotiate necessary financial resources. The inspiration for setting such a financial tool came from the other European regions which applied innovation vouchers as a tool for commencing the collaboration of companies and research institutions. Since the inception of this subsidy program, there have been made several improvements based on the past experience.

- Large businesses have been excluded from the scope of eligible applicants.
- The subsidy for one innovation voucher has been increased from 12,000 euros to 16,000 euros.
- Selection of submitted project proposals has been changed from the drawing lots to the assessment by expert committee.
- The annexes of the application have been supplemented with requirements to compile a project plan.
- There have been made some alternation regarding the condition of innovation voucher award. In the past, the only knowledge providers from the Region were eligible to collaborate; the companies were not limited to apply. Since 2013 it has been the other way around, i.e. the only companies from the Region have been eligible to apply; the knowledge providers have not been limited to collaborate.

All these above mentioned measures are to endorse more effective utilization of public funds in terms of quality of projects.

As for the legal issues, the innovation vouchers have no institute stipulated in the Czech or EU law. It is a specific tool for promoting regional development which is mainly utilized to support activation of cooperation between research institutions and business entities. Voucher provision is governed by EU and Czech law according to the model chosen. Legally it is possible to set these kinds of programs loosely while respecting the law terms and conditions governing the use of public resources for R&D. When chosen a particular model of funding, you have to harmonize it with state aid rules and European legislation in this area and national regulations.

7.2.3. The innovation voucher today

The calls for proposals are opened by the Regional Authority. As mentioned above, the objective of the innovation voucher is to provide financial resource to the companies which are active in R&D and aim to cooperate with universities and other research organizations. The program is funded by the Regional Authority and is opened annually.

Who can apply?

All small and medium-sized legal entities that seek the research assistance of universities or research organizations based in the Moravian-Silezian Region.

How much and what you can get?

The funding can cover the entire process of the development of product or service including testing, designing prototypes or final products as well as creation of market analysis and a business plan.

How much money can I raise?

The maximum funding of one collaboration is 16,000 euros. Given that this is a relatively small amount of funding provided per application and this tool should have more "incentive" effect, it is necessary to minimize the related paperwork for both the applicant from companies as well as knowledge provider (it is one of the main objectives).

Of course, the administrative costs in the view of different models vary and are higher when using funds from the European Union. Nevertheless, the best way how to attract it for applicants is to avoid transferring the administrative burden on applicants and leave it to the provider or the implementing agency. In practice, this means that unlike regular grant applications and the related confirmation and attachments, applications for innovation vouchers are very simple and concise and in many cases the attachments are replaced by statutory declaration. Similarly, the process of implementation has tended to focus on the content, not on the administration process, so the billing is usually very simple as well as final reporting concerning the quality of cooperation. The below mentioned project is a typical sample of innovation voucher result.

Development of interior LED lighting, the company TRIMMER Ltd.

Objective:
- diversification of production program
- involvement of postgraduates from VSB-Technical University of Ostrava.

Development:
- new approach to the philosophy of guiding lumen
- measurement and testing
- development of components
- analysis of the suitability of materials
- production of a functional prototype samples
- preparation of the documentation for the certification.

Result:
- LED interior lights — dim TFN
- maximum efficiency in electrical and lighting parts
- high efficiency, lower energy consumption, longer lifetime.

7.2.4. *The innovation voucher in the future*

The first call of innovation vouchers in the Region was established in 2010. Table 7-1 gives an overview of the number of innovation voucher applications for the different calls since 2010. Generally, this project is to evaluate the following:

- Knowledge providers — they managed to involve prestigious providers of knowledge (e.g. VSB — Technical University of Ostrava and University of Ostrava).
- The project involved 132 applicants (all that applied) and exceeded the amount of funds allocated. This interest might be probably attributed to the quality of participating knowledge providers and balanced set of simple conditions and requirements on applicants.

Table 7-1: Timeline of the innovation voucher applications (source: final meeting innovation voucher beneficiaries).

Year	Subsidy in total	Number of unapproved applications	Number of approved applications	Innovation vouchers			
				Number of applications in total	Subsidy value of unapproved applications	Subsidy value of approved applications	Subsidy value of all applications
2010 (1st call)	400 000 €	0	15	15	0 €	156 480 €	156 480 €
2010 (2nd call)	400 000 €	0	13	13	0 €	130 760 €	130 760 €
2011	400 000 €	3	39	42	72 000 €	391 620 €	463 620 €
2012	400 000 €	37	27	64	512 592 €	391 104 €	903 696 €
2013	400 000 €	26	26	42	358 380 €	390 236 €	748 616 €
CELKEM	2 000 000 €	66	120	186	942 972 €	1 460 200 €	2 403 172 €

- About 50% of applications in 2012 and 2013 were rejected due to excess of demand over supply. The selection has been improving, from drawing lots to expert committee.

It is necessary to prevent the misuse of public funds, but the administration should not be onerous for applicants. Obviously, too general and open system increases the risk of misuse of public resources. In this context, the biggest problems might occur provided that:

- Services are very broadly defined and lack coherence on clear goals — in this case there is no expected impact generated.
- The expected activities and outputs are not realized in the sufficient quality due to acceptance of lower quality service providers. This issue can lead to costly conflicts among small and medium-sized enterprises, service providers and bodies issuing vouchers.
- The scheme is misused by small and medium-sized enterprises and service providers (especially when a small and medium-sized enterprise can act as a service provider) — the higher the value of the voucher, the more problems can emerge.

Hence, the voucher system should be in compliance with risk management, which reflects the various limitations of the system. In general, there is a large positive correlation between the value of the voucher and the amount of restrictions on services that can be supported. The level of risk increases with the adoption of private or foreign unfamiliar service providers. R&D centers in the region are more known and trusted (even at the cost of lower excellence) than foreign institutions without a history of cooperation in the region. Nevertheless, low administrative costs is very important feature of innovation vouchers, therefore the possibilities of risk elimination are rather small and based on the prudence of officers and committee members. Also, the success of the project can be assessed only with the knowledge of such indicators as a scope of ongoing cooperation, the volume of other contracts, changes in attitude, etc. However, these indicators can be distorted by inappropriate conditions set in future calls (e.g. allowing repeated submission of projects between the same applicant and knowledge provider), implementation of vouchers focused on a similar target group (e.g. the applicant may be the regional company which has received a subsidy for cooperation with a regional university but also applies in other calls in different regions where they can get more vouchers for another project with the same knowledge provider). In this case, it is a form of financial support for a specific project and not a means of promoting the establishment and testing of cooperation. Repeated cooperation does not necessarily need to be motivated by the experience of the previous project and verified quality but it does need to be driven by the possibility of obtaining a grant for the project.

Another serious concern regarding innovation voucher is their short-term effect. The vouchers only facilitate one-off and subsidized industry–university cooperation, leaving unaltered the long-term attitude of SMEs towards R&D and innovation. However, it is unclear whether this corresponds to a real general pattern or if it is linked to the pattern of innovation activity in smaller firms, which tend to be relatively spasmodic. If so, the impact of a voucher on SMEs' further engagement with universities can only be measured in the very long term.

The last, but in no means least, limitation of innovation vouchers is technology lock-ins. If the scheme provides for the knowledge institution to be from the same country or region of the company, this can limit the search patterns of SMEs and their ability to find an effective solution to any technological problems.

7.3. Implementing best practice

The innovation voucher is a financial instrument to promote cooperation test (usually called initial cooperation) between business and academic sectors (which we use here in a broader sense to mean universities, research and development institutions — the "knowledge providers"). For these funds, businesses the knowledge providers buy services or knowledge, especially in terms of addressing specific innovative projects, measurement, analysis, studies and final reports, proposals and prototypes, etc. The M-S Region innovation voucher can be summarized as follows:

- subsidy worth up to 16,000 euros covering up to 80% of the supported project
- intended for a purchase of knowledge from all around the world (since 2013)
- for companies coming from M-S Region (since 2013)
- with minimal administrative burden
- an effective tool for promoting technology transfer.

The subsidy is usually provided to the voucher recipient in the de-minimize[3] regime. The subsidy mustn't be used to cover the same cost of a project financed completely or partially from other public grant schemes.

7.4. Implementing best practices in your region

7.4.1. *Implications for the implementation body (e.g. Agency, TTO)*

The innovation voucher contract is intended for the provision of services under the contract between the applicant and the provider of knowledge (Figure 7-1). The contract shall be specified: the subject of performance, cost, schedule, and professional competence of knowledge provider (especially people and equipment, or more precisely technologies), the existence of R&D results, methods of verifying achievements, rules and obligations of the result utilization.

Application of the voucher should include information regarding: general objective of the project (mission, why is the project carried

---

[3] De minimis rule — subsidies which are below a certain minimum level don't have to be notified and approved by the European Commission. This applies to grants which are given to individual companies by the state or by public authorities within the current fiscal year and the last two calendar years with a value of up to 200.000 euro in total (100.000 euro in the field of road transport sector). The commission is assuming that these smaller subsidies will have no significant impact on trade and competition between the member states.

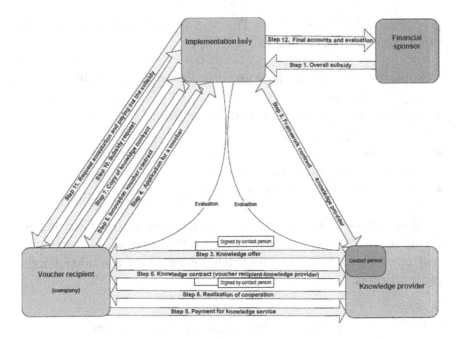

Figure 7-1: Innovation voucher diagram. Source: JIC, 2013.

out, what problem will be solved, etc.); specific project objectives (what needs to be done specifically in order to solve the problem); target groups; benefits for the target groups; justification of need for the project; key project activities; project schedule; project risks; project indicator monitoring; implementation team; sustainability of the project; level of product innovation, process or service and project's contribution to the competitiveness of the applicant.

As for the costs, the only applicable costs are the non-investment costs that the applicant pays to the knowledge provider with whom he contract for work or for the supply of knowledge. The associated costs are: development of product/process/service; testing measurements; feasibility studies; access to research facilities; prototyping; analysis of the suitability of the material; product design; creation of business plans for innovative product; assessing the economic impact; market analysis/marketing strategy; innovation/ technology audit; setting up a new business model or management of the company and optimization of operational business processes.

Formal and content checks of the incoming applications is carried out by a three-member committee of experts who check the technical feasibility of the project (level of innovation), marketing (the applicability of the product on the market, providing sales) and economic aspects (assessment of financial resources, profitability, payback period, sustainability, etc.). The committee of experts has the right to exclude any application that doesn't comply with the objectives of the program.

While judging the eligibility of the applications, the committee of experts will look for answers to the following questions.

Formal eligibility

- Has the application been well filled-in and submitted? Does it fulfill all formal criteria?
- Is the identification data of the applicant correct?
- Is the applicant eligible for asking for a voucher?
- Was the exclusion of the conflict of interest confirmed?

Content eligibility

- Does the application describe the subject of proposed cooperation concretely and comprehensibly enough?
- Does the application contain only eligible activities?
- Is the maximum cost of the proposed collaboration 16,000 euros total?
- Has the applicant described the way of using the purchased knowledge sufficiently?
- Has the applicant specified the expected benefit for strengthening its competitiveness sufficiently?
- Will the purchased knowledge be usable by the applicant in a sustainable way? (The voucher shouldn't be a subcontract to a one-time contract for a third party.)
- Are the budgeted costs adequate?

- According to the application, is it possible to assume that the knowledge is new for the applicant and that it is not commonly available?
- Is the application sufficient in order to persuade the committee of experts that the proposed collaboration complies with the project objectives?

7.4.2. Implications for universities/research organizations

The universities usually represent the knowledge provider, so their role is clear. In some countries there were attempts to have universities act as the delivery agency. However, this practice has the potential to cause conflict of interest and shifts too heavy a burden onto the university management. Experience suggests that public authorities and agencies are best suited for the brokering role that a voucher scheme requires. However, what should be mentioned are the benefits for the universities and other research organizations, which are as follows:

- Opportunities for cooperation and preparation of high-quality larger projects for national programs.
- Finding new industrial partners. Regional research institutions have so far obtained tens of new contacts with companies.
- Thanks to the vouchers researchers have specific feedback regarding the need for and value of their activities. This helps them e.g. with pricing of their service. At the same time, company demand provides them with important information for the direction of their research.

7.4.3. Implications for policy makers

The innovation voucher program has been a tool for policy makers to support what have often been first innovation activities in SMEs. New and better approaches should be developed and tested through the regional pilot projects and rolled out as widely as possible. Certainly, the policy makers aim to improve SMEs innovation performance, however, they should consider how to approach the voucher system within the overall policy mix.

It is up to policy makers to decide how much funding will be allocated and to decide on the best suited level of governance for this type of innovation support measure based on the overall objective of the scheme, the desirability of having or not a regional specialization, the regional competences and the governance framework in general.

The obligation of the politicians is to ensure that the research they support is of good quality and has tangible benefits for business. But the responsibility is wider than the translation of research into application. In an environment where the future health of our economy depends on our research output and our ability to capitalize upon this, the decision about which fields to support and how much money is allocated to them and, by inference, which fields not to support, carries a heavy burden of responsibility. They need to be informed by leading-edge thinking and in-depth experience and then they can fulfill that role.

7.5. Future opportunities

Further improvement measures should be based on better project assessment, selection and evaluation of the long-term objectives. Regarding the assessment of the applications, some measures have already been put in place concerning the selection of submitted project proposals (from drawing lots to expert committee, requirement to submit project plans as an obligatory attachment, etc.). Nonetheless, there is still some room to improve the assessing procedure by e.g. reducing the amount of subsidy per voucher (from 80% to 50%) in order to stimulate the applicants that are really interested into the university–company collaboration. Also the evaluation of the program results should be carried out on the regular basis with an objective to assess not only the number of approved applications and exhausted funds but also to answer any questions concerning the benefits (new contacts, solutions, markets etc.) brought to the collaboration after the end of the project and the impact of this on the sustainability of other R&D projects.

In summary, the current system of the innovation voucher can be improved in the following ways:

- Better risk management measures.
- Reduction of low quality projects thanks to a better selection procedure.
- Better evaluative analyses (especially short- and long-term behavior) on the adoption of efficient measures in the future.

7.6. Best practices

- Innovation vouchers offer the opportunity for companies and universities to work together for the first time, which can result in new long-term collaborations.
- The program is embedded in the Regional Innovation Strategy implementation scheme.
- Works best with an application process that has minimal administrative burden.

7.7. References

Papers, reports, books, conference presentations and websites

Moravian-Silesian Region (2014). Conditions of the subsidy programs for support of business in Moravian-Silesian Region. Retrieved from: http://verejna-sprava.kr-moravskoslezsky.cz/cz/uredni_deska/podminky-dotacniho-programu-podpora-podnikani-v-moravskoslezskem-kraji-2013-30446/. Accessed on: 30.06.2014.

RDAO (Regional Development Agency Ostrava) (2012). Regional Innovation Strategy of the Moravian-Silesian Region 2010–2020. Retrieved from: http://www.rismsk.cz/en/download/. Accessed on: 30.06.2014.

The South Moravian Innovation Centre (JIC), (2013). "Project manual — Inovační vouchery".

Stejskal, J. and Matatkova, K. (2012). Innovation Vouchers as a Suitable Instrument for Effective Public Support of Innovations by Local Public Administration. *International Journal of Systems Applications, Engineering and Development*, 6(5), 333–341.

Case study interviews

Regular quarterly interview with the representatives of the Department of Regional Development and Tourism at the Regional Authority.
Final meeting with representatives of the beneficiaries of innovation vouchers.

8

CASE STUDY 8: Public Funds for Patenting, Valorization and Science–Industry Collaboration

Magdalena Powierża and Piotr Potepa
(International Institute of Molecular and Cell Biology, Poland)

8.1. Setting the scene

Poland is an excellent place for performing research and development. The economic transformation gave rise to changes that made Poland a modern, stable country with high investment potential. In the heart of Poland is the Mazovia region which occupies 35,558 km^2 and has 5,268,660 inhabitants. It is worth noticing that the urbanization factor amounts to 64.7%. Mazovia has the highest income per capita among all Polish provinces, which constitutes around 85% of the average income in the European Union. The Polish scientific and academic potential is based on 460 universities and higher education schools, 2 million students and 200 research and development centers with nearly 100,000 research and scientific staff.

The Polish government has realized that the country cannot sustain its good economic results without a strong position in science. There is an abundant flow of money to stimulate science–industry collaboration — an area that has been vastly neglected for years. The money comes mostly from the EU funds and the state. Poland joined the EU on May 1, 2004 and subsequently became eligible for support from the EU Structural and Cohesion Funds. The primary objective of these funds is assistance in reducing the development disparities between regions in order to strengthen the economic and social cohesion. The program Innovative Economy is one of six national programs under the National Strategic Reference Framework, which are co-financed by EU resources. This program is directed to all entrepreneurs who want to implement innovative projects connected with research and development, modern technologies, investments of high importance for the economy or implementation and use of information and communication technologies. This program consists of nine priority axes which are divided into detailed measures. Support may be granted under those individual measures. Every priority axis is focused on the support of particular types of projects and therefore implements particular specific objectives of the program (European Funds Portal, 2014).

The main administrator of the national resources allocated to support R&D is the National Centre for Research and Development (NCRD), which offers a number of different opportunities for co-financing R&D projects. The NCRD is the implementing agency of the Minister of Science and Higher Education. It is in charge of the performance of the tasks within the area of national science, science and technology and innovation policies. It was the first entity of this type, created as the platform of an effective dialogue between the scientific and business communities. The science reform adopted in autumn 2010 gave the NCRD more freedom to manage its financial assets, within the scope of a strategic research program. It has the function of the Mediation Institution in three operational programs: Human Capital, Innovative Economy and Infrastructure and Environment. The Centre became one of the greatest innovation centers in Poland. The activity of the Centre is funded by the national treasury and the European Union (NCRD, 2014).

The NCRD offers a number of different opportunities for co-financing R&D projects. Thanks to this support it is possible to fund various initiatives, including research-related ones. Implementation of R&D projects and creation of research and development centers plays a vital role in building a knowledge- and innovation-based economy, therefore, allocation of EU funds for 2007–2013 included support to create new/expand the existing R&D centers. The main program responsible in this matter is Innovative Economy program.

The upcoming financial perspective of the European Union for 2014–2020 will put much more emphasis on the implementation of the Strategy "Europe 2020", supporting smart growth through the development of a knowledge- and innovation-based economy, sustainable growth by promoting efficient use of resources, increased competition or inclusive growth by promoting high levels of employment and ensuring social and territorial cohesion. It is not yet known how exactly it will look like with patenting and science-industry collaboration. To sum up, most of the funds for R&D projects in Poland come from the European Union and NCRD.

8.2. Identifying best practices

This case study introduces the diverse system of public support programs for technology transfer in Poland. The following part describes the main public programs and measures.

8.2.1. *Programs offered by National Centre for Research and Development*

8.2.1.1. *Innovativeness creator*

The development of a contemporary economy calls for an intense cooperation of science and economy. Creativity and innovation are important factors enhancing the competitiveness of any economy. The Innovativeness Creator program enables the implementation of projects encouraging broadly understood research and commercialization. It is addressed to public research entities (e.g. universities, institutes) with the purpose of stimulating actions taken by them to commercialize their scientific knowledge and know-how through:

- development of public R&D commercialization systems
- educational and training activity related to commercialization of scientific knowledge and know-how

- promoting entrepreneurship among students, graduates, university staff and researchers.

The program is expected to:

- contribute to the increase in the number of commercialized technologies and solutions
- increase the number of commercialized innovative technologies and solutions
- develop a network of units supporting entrepreneurship among scientists
- raise the efficiency and effectiveness of actions taken by entities promoting entrepreneurship
- promote information on opportunities and benefits of establishing companies on the grounds of intellectual property rights
- promote best practices for the commercialization of knowledge and know-how and for copyright protection.

The costs are eligible if they are related to database systems (commercialization of new technologies), procedures for the management of intellectual property, consultancy and training, participation in exhibitions and fairs (science-to-business) or for info-promo activities (NCRD, 2014).

8.2.1.2. *Patent plus*

The PATENT PLUS program is a financial support instrument with the main purpose of increasing the intellectual property management effectiveness via patenting. PATENT PLUS is also addressed to public research entities to encourage both scientists and management of research entities to file for patent protection of their R&D results. Its main objective is to increase the number of patent applications leading to a better protection of industrial property rights in Poland by way of co-funding or refunding of costs incurred by submission of a patent application. Owing to this program research entities may develop a properly balanced portfolio of R&D projects with IPR in place, which will attract business partners, and result in intensified commercialization of developed inventions. Eligible costs are related to the preparation of the patent application, effectiveness analysis, commercialization strategy, translation, feasibility study and PCT procedure (NCRD, 2014).

8.2.1.3. Applied research program

Applied Research Program is a horizontal program to support the science and business sector in applied research in various fields of science and industry. The program funds:

- research and development institutes (units) or scientific consortia that aim to acquire knowledge in a particular field of science with practical potential
- industrial research and technical feasibility studies of consortia and research industrial centers.

Eligible costs are those related to laboratory equipment, rental of buildings and research services (NCRD, 2014).

8.2.1.4. SPIN-TECH Special Purpose Vehicle (SPV)

SPIN-TECH is a new Program of the National Research and Development Centre to support operational activities of research units called Special Purpose Vehicles (SPV) that are established to commercialize research results. The main objective of the program is to increase the research results' commercialization through SPVs acting as intermediaries between the public R&D sector and the market. The program is also designed to intensify technology transfer from science to the economy and to accelerate development of entrepreneurship among scientists through setting up spin-off companies by academics and students. The SPIN-TECH program funding is devoted to (NCRD, 2014):

- identification and assessment of the commercial potential of R&D results
- assessment of possible commercialization means and selection of the optimal solution

- initiation of SPV operation, e.g.: development of legal documentation, organizational work and business plan preparation
- business model and internal procedures preparation
- potential clients identification.

8.2.1.5. Operational Programme Innovative Economy (OP IE)

The Operational Programme Innovative Economy (OP IE) offers opportunities for financing scientific research and technology transfer collaborations with enterprises which operate in Poland, under two Priority Axes:

- Priority Axis 1. Research and development of new technologies.
- Priority Axis 4. Investments in innovative undertakings.

The intended effect of activities in both Priority Axes is increased significance of science in the economy through strengthening R&D work in the areas considered to be of key importance for the social and economic development of the country and improving the level of innovativeness of enterprises. Public research institutions may apply for the support only if they are in a working consortium with a company.

One of the measures in Priority Axis 1 is the support for R&D projects for entrepreneurs carried out by scientific entities (OP IE 1.3). The objective of the measure is to acquire and increase the use of new solutions essential for developing the economy, for improving the competitive position of entrepreneurs and for developing Polish society. Due to insufficient use of R&D results from Polish science institutions in economic practice, it is essential to develop instruments that increase the supply of new, innovative solutions useful for entrepreneurs. These types of instruments are development projects worked out by scientific units to bridge the gap between industrial research and later implementations. These projects aim at practical application for the needs of the sector/branch of the economy. They can also be of high importance to the society because their results should contribute to solving the most current social problems. The financing of such projects enables more efficient and faster implementation of goal-oriented projects which help to build the competitive position of Polish R&D organizations on the market. This program finances the costs of research centers and scientific–industrial consortia that are related to laboratory equipment, services research, analysis, legal expertise and project promotion.

Tightening of cooperation between the science sector and enterprise requires taking up actions supporting protection, commercialization and disseminating results of R&D works. Within the presented measure parts of the costs connected with gaining legal protection of R&D results are financed. This includes patent proceedings, project promotion, legal expertise and delegations.

Potential beneficiaries can also get support under two other measures of the Operational Programme Innovative Economy: Measure 1.4. Support for goal-oriented projects and Measure 4.1 Support for the implementation of R&D results. In contrast to OP IE 1.3, these measures are addressed more to companies than to public R&D institutions. Applicants file one application covering two stages: the research and its implementation. The condition for obtaining investment support for the second part of the project is to successfully complete the first stage. This means that the prototype developed should offer a chance of market success for the new product.

Support under OP IE, Measures 1.4 and 4.1, is available to both micro, small, medium-sized and large enterprises established in Poland. Entrepreneurs can either have an R&D section in-house or work together with a contractor who can be a scientific institute, a scientific network or another entrepreneur who has the required facilities. They can also do the work by themselves if they have the required infrastructure and other necessary resources (PAED, 2014).

The measures can be used to fund technological or organizational projects (industrial research and development work). The financial support can cover the purchase of fixed or intangible assets together with the necessary consultancy required for the implementation of the results of R&D work done during stage one of the project (with consultancy being eligible for co-financing only in the case of SMEs). It is worth mentioning that Measure 1.4 covers the co-financing of expenses until the development of the prototype while Measure 4.1 is a continuation of the project where more attention goes to the implementation of R&D work. This means that entrepreneurs should plan the research and implementation stages of their projects in advance.

8.3. Implementing best practices in your region

To implement this structure of programs, one needs to have a government which is capable of aligning the different programs with a clear and comprehensive strategy focused on technology transfer. Furthermore, it should be easy for all stakeholders in the technology transfer story to identify these programs and apply without too much administrative work. Below, we present how the application and monitoring process works for our programs.

8.3.1. *Application process*

The projects are selected as a result of project contest announced and conducted by particular Implementing Authorities (sometimes 2nd level Intermediate Bodies) which are responsible for implementation of a given measure. Selection of these projects is performed with respect to the principle of disclosure and access to information according to the criteria of project selection adopted by the Monitoring Committee. The process of project selection consists of the following stages:

- call for proposal
- submission of projects
- formal evaluation and content-related evaluation of applications
- publication of the contest results
- review procedures (if needed)

- signing contracts on financing projects
- registration of documents in the information system, according to separate provisions in areas concerned (the first registration after the formal evaluation of the application for support).

8.3.2. *Monitoring of projects*

The beneficiary is obliged to submit the application electronically (using special generator requests for payment) and printed on paper. The application may relate to payment, reimbursement or a settlement advance. The application will sometimes be accompanied by an additional table; summary of the cumulative expenditure by categories within each payment applications aimed to analyze the correctness of the implementation of the agreement funding. The beneficiary is required to apply for payment on a quarterly or half-year basis (which depends on the program).

8.4. Best practices

- Big variety of programs addressed both to public and private entities, and often to consortia including the two parties.
- Funding for different kinds of activities.
- Offer training and support for beneficiaries to learn how to apply for money, how to fill in request for payment, etc.

8.5. Future opportunities

- The government aims to decrease the bureaucracy and to simplify the application and speed up the approval so that there are fewer delays in signing contracts and paying beneficiaries.
- There are some problems with transferring money from one activity to other. The aim is to make the budgets less rigid so that there is more flexibility to allocate the different budgets.
- The different implementing agencies all have different rules of monitoring projects. Future efforts should concentrate on aligning these rules.
- Investments in co-financed projects should be supported by a more careful analysis of the market.

8.6. References

Papers, reports, books, conference presentations and websites

European Funds Portal (2014). Retrieved from: http://www.funduszeeuropejskie.gov.pl. Accessed on 05.07.2014.
NCRD (National Center for Research and Development) (2014). Retrieved from: http://www.ncbir.pl/. Accessed on 05.07.2014.
PAED (Polish Agency for Enterprise Development) (2014). Retrieved from: http://en.parp.gov.pl/. Accessed on 05.07.2014.

Further reading

Gwiazda M., Mazurek B., Zalewska A., Rebkowiec G. & Leśniewski Ł. (2012). "R&D market in Poland — support for research and development activity of enterprises", Tax and, Polish Information and Foreign Investment Agency, National Centre for Research and Development.
Ministry of Regional Development (2007). Operational Programme, Innovative Economy 2007–2013. Retrieved from: http://www.poig.gov.pl/English/Documents_POIG/Documents/innowacyjnagospodarkaang_X_2007.pdf. Accessed on 06.07.2014.
Ministry of Regional Development (2009). Detailed description of the priorities of Operational Programme Innovative Economy 2007–2013. Retrieved from: http://www.poig.gov.pl/English/Documents_POIG/Documents/Szczegolec_POIG_29012009_EN.pdf. Accessed on 06.07.2014.
National Information Processing Institute (2014). Retrieved from: http://www.opi.org.pl. Accessed on 06.07.2014.
Operational Programme Innovative Economy (2014). Retrieved from: http://www.poig.gov.pl. Accessed on 06.07.2014.

SECTION 3: INCUBATORS

Introduction

The dominant way in which technology has been traditionally transferred from the university sector to the private sector is through technology licensing (Siegel *et al.*, 2003). License agreements allow the academics to pursue their research without having to commit large amounts of time to commercial matters. However, in the past decade there has been a substantial rise in the creation of university spin-outs as an attractive alternative to licensing technologies (Lockett & Wright, 2005). As research on knowledge spillover and organizational learning shows that continuous interactions among creators, appropriators and consumers of technology accelerate the richness and reach of knowledge and discoveries (Agrawal & Henderson, 2002), universities need to organize the process of accelerating technology spillover and innovation in universities. Universities developed incubators, sciences parks and/or accelerators to facilitate the early stages of spin-outs and create interactions between inventors and the industry. Especially in environments with less demand for innovation, characterized by a weak entrepreneurial community and few other key resources, universities may need to play a more proactive incubation role (Clarysse *et al.*, 2005). In this section, we focus on best practices in setting up university incubators, science parks or accelerator programs.

University incubators are usually physical spaces attached to the university to help commercialize its own spin-outs and foster business ideas from its network in exchange for a monthly rental fee. Its main characteristics are the provision of physical office space, mentorship network, informal event programs, consulting services, investor exposure and public funding links (Salido *et al.*, 2013). The main goal is to produce successful businesses that will leave the program financially viable and freestanding. New ventures typically can remain in such an incubator for several years, depending on their needs and the focus of the incubator (Lewis *et al.*, 2011).

University incubators can be seen as one of the four types of incubator models (Barbero *et al.*, 2014). First, they identify business innovation centers which are established to stimulate regional economic development. These have been set up with the support of policy makers to generate employment in the region (Aernoudt, 2004). For instance, it provides seed capital to invest in IP (Clarysse *et al.*, 2005; Clarysse & Bruneel, 2007) and shared laboratory equipment to develop the technology further to the market (Wright *et al.*, 2007). The second type is a basic research incubator built around research institutes seeking to valorize research output. In comparison to the previous type, this research incubator focuses on a specific sector that is strategic for the naturally formed cluster and region it is created in (Aernoudt, 2004). This incubator puts effort on connecting the new and established ventures within the sector cluster. Consequently, by operating in clusters and promoting bilateral collaboration, it generates more innovations in comparison to university and region development incubators (Oakey, 2007). The third type of incubator is a privately funded organization with a high-risk investment model for the support of high-potential new ventures (Grimaldi & Grandi, 2005). This type of incubator originates from the venture capital and corporate industry. The fourth type is university incubators which are created to facilitate technology commercialization from research departments to the industry. These have been set up by universities to nurture academic spin-offs (Von Zedtwitz & Grimaldi, 2006). This incubator is distinctive in typically having a technology focus and specializing in a number of aspects that are important to spin-offs. Despite the differences between the incubators, the contemporary incubators each offer at least four of the five following services: (1) access to physical resources; (2) office support services; (3) access to capital; (4) process support such as mentoring; and (5) networking services (Von Zedtwitz, 2003).

The accelerator model, which became globally famous with Y-Combinator in the US, has spread to many European hubs. Although accelerators provide mentoring and networking to their portfolio ventures, they also differ substantially from them. First, they do not provide physical resources or office support services. Second, they are less focused on venture capitalists as a next step of finance, but are more closely connected to business angels and small-scale individual investors. One of the reasons for this difference is that their focus is not on capital-intensive start-ups nor on technology oriented spin-offs. At the time when incubators emerged, many of the innovative ventures were active in sectors such as biotechnology, microelectronics and electrical equipment which are typically capital intensive (Wright *et al.*, 2007). Support is time-limited and comprises events and intensive mentoring and the program itself tends to be organized in batches of start-ups beginning at the same time. The number of European accelerators and incubators has increased dramatically since the start of the financial crisis. Between 2007 and 2013, the number has risen nearly 400% (Salido *et al.*, 2013).

Besides incubators and accelerators, technology transfer can also be facilitated by science parks. A science park is an organization managed by specialized professionals, whose main aim is to increase the wealth of its community by promoting the culture of innovation and the competitiveness of its associated businesses and knowledge-based institutions (Ratinho & Henriques, 2010). Supporting young technology-based firms to establish and flourish as well as attracting anchor firms to a given location is also often among their objectives.

In what follows, we discuss the best practices in founding and developing incubators (Case Study 9) and accelerators (Case Study 10). Case Study 9 explains the foundation of the Imperial College bioincubator and underlines the importance of the incubator's design to facilitate network activities and create an entrepreneurial community.

The close collaboration with Imperial Innovations (TTO) is also crucial as they follow a high-quality screening process. While the Imperial College bioincubator is very technology push oriented, Case Study 10 discusses the activities of Idea lab, an idea development program to offer students of all faculties an opportunity to implement their most (positively) outrageous ideas, try out new concepts and build prototypes in multidisciplinary teams. This program support ideas and knowledge transfer from students and shows a lot of similarities with the accelerator program. The success of many US start-ups founded by students pointed out that a lot of interesting ideas are developed by non-academic staff and are not high-tech and sometimes not protectable. Idea lab is an answer to this trend. In line with this, Wright *et al.* (2009) argue that "business schools through faculty and experienced MBA students may have an important role in academic entrepreneurship through the development of internal university processes that promote rather than hinder internal knowledge flows between business schools, TTOs and science departments".

To summarize, this section shows best practices in three different forms of incubation and underlines the importance for a university to protect, support and facilitate the early stages of their spin-outs. Universities should not only focus on inventions developed by academics but should also support and facilitate the transfer of ideas coming from different types of students active in different faculties.

References

Aernoudt, R. (2004). Incubators: Tool for entrepreneurship? *Small Business Economics,* 23(2), 127–135.

Agrawal, A. & Henderson, R. (2002). Putting patents in context: Exploring knowledge transfer from MIT. *Management Science,* 48(1), 44–60.

Barbero, J.L., Casillas, J.C., Wright, M. & Garcia, A.R. (2014). Do different types of incubators produce different types of innovations? *The Journal of Technology Transfer,* 39(2), 151–168.

Clarysse, B., Wright, M., Lockett, A., Van de Velde, E. & Vohora, A. (2005). Spinning out new ventures: A typology of incubation strategies from European research institutions. *Journal of Business Venturing,* 20(2), 183–216.

Clarysse, B. & Bruneel, J. (2007). Nurturing and growing innovative startups: the role of policy as integrator. *R&D Management*, 37(2), 139–149.

Grimaldi, R. & Grandi, A. (2005). Business incubators and new venture creation: an assessment of incubating models. *Technovation*, 25(2), 111–121.

Lewis, D.A., Harper-Anderson, A. & Molnar, L.A. (2011). Incubating Success. Incubation Best Practices That Lead to Successful New Ventures. Report US Department of Commerce Economic Development Administration (EDA)

Lockett, A. & Wright, M. (2005). Resources, capabilities, risk capital and the creation of university spin-out companies. *Research Policy*, 34(7), 1043–1057.

Oakey, R. (2007). Clustering and the R&D management of high-technology small firms: in theory and practice. *R&D Management*, 37(3), 237–248.

Ratinho, T. & Henriques, E. (2010). The role of science parks and business incubators in converging countries: Evidence from Portugal. *Technovation*, 30(4), 278–290.

Salido, E., Sabas, M. & Freixas, P. (2013). *The Accelerator and Incubator Ecosystem in Europe*. S. E. Initiative, Telefonica: 24.

Siegel, D.S., Waldman, D. & Link, A. (2003). Assessing the impact of organizational practices on the relative productivity of university technology transfer offices: an exploratory study. *Research Policy*, 32(1), 27–48.

Von Zedtwitz, M.V. (2003). Classification and management of incubators: aligning strategic objectives and competitive scope for new business facilitation. *International Journal of Entrepreneurship and Innovation Management*, 3(1), 176–196.

Von Zedtwitz, M. & Grimaldi, R. (2006). Are service profiles incubator-specific? Results from an empirical investigation in Italy. *The Journal of Technology Transfer*, 31(4), 459–468.

Wright, M., Clarysse, B., Mustar, P. & Lockett, A. (2007). *Academic Entrepreneurship in Europe*, Edward Elgar Publishing.

Wright, M., Piva, E., Mosey, S. & Lockett, A. (2009). Academic entrepreneurship and business schools. *The Journal of Technology Transfer*, 34(6), 560–587.

9

CASE STUDY 9: The Imperial Bioincubator

Robin De Cock
(Ghent University, Belgium; Imperial College London, UK)

9.1. Setting the scene

In 2010, 942 companies were involved in bioscience in the UK, including 345 companies directly involved in the development, manufacturing or selling of therapeutic products. The UK Bioscience sector provides highly skilled employment for 36,000 people. In 2010 turnover totaled £5.5 billion in 2010, an increase of 18% in just one year. Biopharmaceuticals invest more in R&D than any other sector in the UK, representing over 30% of our total R&D spend in the UK. According to PricewaterhouseCoopers (PwC), the global market for biopharmaceuticals grew at a rate of more than 20% between 2002 and 2007. The value of biological medicines in development in the UK represents around £24 billion in a global pipeline which is estimated to be worth over £200 billion. Seventy percent of all biologics in development in the UK are in the later stages (Phase III and pre-registration). Biological medicines in development in the UK are valued at around £15 billion. Biotech has created more than 200 new therapies and vaccines, including products to treat cancer, diabetes, HIV/AIDS and autoimmune disorders. More than 325,000,000 patients worldwide have benefited from approved biotech medicines (London Medicine, 2013).

The biotech sector in London is a rapid-growth sector. Some sources (former London biotech network, now London Medicine) speak of 84 biotech companies in London in 2005, which is an increase of about 40% since 2001, showing a significant spurt in growth. The majority of these new companies are university spin-outs. Most of the biotech companies in London are still at an early stage; over 70% of the companies are start-ups employing fewer than 10 people. Then there is a smaller group (22%) of quickly growing companies as well as a much smaller group (8%) of more "mature" companies, employing over 30 people. Currently there are eight publicly listed companies based in London with a range of market capitalization values up to around £40 million for the largest companies, for instance Antisoma. The companies' main focus areas are very varied, but new cancer and cardiac therapies are perhaps the most prevalent. Other areas range from inflammatory diseases, to bio-informatics, genomics and tissue engineering. The majority of the companies are focused on new medical therapies, stemming from their origins in the medical schools in London (London Medicine, 2013).

The companies are clustered close to the academic research centers in central London and also predominately towards the west of London. This to a large extent reflects the origins of biotechnology companies spinning out from the medical schools and universities of central London and migrating westwards as they grow. London has a natural cluster of bioscience research within its medical schools, universities and hospitals — in fact probably the largest in Europe. The spin-out companies naturally like to remain geographically close to their academic centers, as they retain links as they grow. Most of the biotechnology companies in London have spun out of Imperial College, King's, Guy's & St Thomas Medical School and University College London, although there are some spin outs from St George's Medical School and Queen Mary, University of London (London Medicine, 2013).

Imperial College London is located at the heart of London city and is a science-based institution with a reputation for excellence in teaching and research. Imperial College currently has the following three constituent faculties: Imperial College Faculty of Engineering, Imperial College Faculty of Medicine and Imperial College Faculty of Natural Sciences. The Imperial College Business School exists as an academic unit outside of the faculty structure. The Technology Transfer Office is called Imperial Innovations. Imperial Innovations was founded in 1986 as the technology transfer office for Imperial College London, to protect and exploit commercial opportunities arising from research undertaken at the College. In 1997, the Group became a wholly owned subsidiary of Imperial College London and in 2006 was registered on the Alternative Investment Market of the London Stock Exchange, becoming the first UK University commercialization company to do so (Imperial Innovations, 2013).

Innovations has a Technology Pipeline Agreement with Imperial College London which extends until 2020, under which it acts as the Technology Transfer Office for Imperial College London. The Group also acts as the Technology Transfer Office for select NHS Trusts linked to Imperial College London, including Imperial College NHS and North West London Hospital Trusts. Following fundraising in January 2011, the Group has invested larger amounts and maintained its involvement for longer in the most promising opportunities within its portfolio of spin-out companies, with the intention of maximizing exit values. In addition, the Group has made a number of investments in opportunities arising from intellectual property developed at or associated with Cambridge University, Oxford University and University College London, through its relationships with Cambridge Enterprise, Oxford Spin-out Equity Management and UCL Business. These universities are the top four research-intensive universities in Europe with a research income of over £1.2 billion per annum (Imperial Innovations, 2013).

9.2. Identifying best practices

Spinning out new ventures from universities is a long-established phenomenon. During the last decennia, universities have devised more and more proactive policies to stimulate the commercial exploitation of public research through spin-outs (Clarysse et al., 2005). One way to support spin-outs is to develop and install a bioincubator. An incubator provides flexible offices and laboratory spaces which enable life science companies to conduct their research and business in the same location. There are many benefits to life science companies being co-located at bioincubators. First, companies within bioincubators are almost twice as likely to secure investment and the amount they receive is more than double that of non-bioincubator companies. Second, researchers can easily share knowledge and ideas and companies are plugged into networks of experts that can support their business and help them grow. Innovation is further stimulated through the association of many bioincubators with universities giving the life science industry greater access to their research base.

9.2.1. Incubators at Imperial College

Imperial College spin-outs are very well supported when it comes to funding, incubation and office space/infrastructure. The Imperial College Incubator, founded in 2005, is a 24,000 square foot building and was jointly funded with £4 million from Imperial College and £3 million from the London Development Agency (LDA was the Regional Development Agency for Greater London which is now GLE).

> *The London development agency was running a project that recognizes that there was a lack of commercial laboratory space within London and they decided to invest in three bioincubators.* (Director of Imperial College bioincubator, December 2012)

Situated at the campus of Imperial College London, early stage businesses are able to take advantage of being in close proximity to scientists and experts in technology. The Incubator's central location in London also offers easy access to customers, sources of finance and commercial partners. So whether you are an established business, a start-up, or you simply need a virtual office, the bioincubator offers leases to suit you and your company, giving your business the freedom to develop. For business looking for a base in London or for companies expanding into a new market, a virtual office is an easy, cost-effective way of boosting your company's image by having a prestigious business address.

> *Before the bioincubator, entrepreneurs/academics just used the facilities at the research department or moved to a science park outside London. The bioincubator really allowed Imperial to keep their spin-offs close. Academics can also pursue more easily their academic work while supporting the spin-off.* (Director of Imperial College bioincubator, December 2012)

The Imperial facility consists of wet labs, private offices, and several meeting rooms and provides the tenants with a range of research resources, business and support services. The wet labs are equipped to Category 2 level laboratory classification system of physical containment. Biotech spin-outs generally stay in the Incubator for three to four years and pay the market price but the Incubator is quite flexible when spin-outs have difficulties in paying rent. The incubator space per company ranges from 40 to 95 m². The Imperial College Incubator is generally run at or near capacity, and is currently hosting around 25 companies.

> *The bioincubator started with 40 to 60% occupancy. That is not too bad. We reach full capacity around 2009–2010. We have never been under an occupancy of 87%, which is pretty good for an incubator.* (Hewson, 2012)

The BioIncubator has a team of three but can rely on the strategic, management and mentoring advice of Imperial Innovations. The Incubator also offers a professional answering service, mail forwarding service and hot desks. Figure 9-1 shows how the bioincubator is implemented in the Imperial College network. The Incubator building is owned by the College. It is leased to Imperial Incubator Limited who leases space through the TTOs (Imperial Innovations).

Companies cannot just enter the bioincubator. Before that they have to go through a screening process as the director of the bioincubator explains:

> *Companies have to apply by using the application form. We do a due diligence on each company to make sure that they are a benefited company*

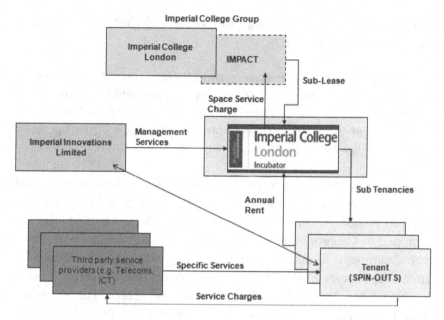

Figure 9-1: Bioincubator in the Imperial College network.

and then we also look into their financials. Can they afford space? We look at their business plan to see if the company is going to complement the existing companies in the incubator. Does it meet our limits of being a scientific company...doing work of scientific interest? If that is ok, then the director of the bioincubator recommends the company to the board of directors. They have the final say on whether a company will join the bioincubator or not. The board of directors includes...people [only] from Imperial College. (Hewson, 2012)

Besides the BioIncubator, the college also has incubation structures for new interdisciplinary businesses for specific key industries as software, healthcare and design, architecture and visual communication (The Health and Care Infrastructure Research and Innovation Centre, design London, Digital economy lab).

Finally, Imperial College is building a new campus (92.000 m²) at the Imperial West site in West London. One of the first buildings, the £150 million Research and Translation Hub (42,000 m²), will be Imperial West's centerpiece with space for 1,000 researchers alongside

Figure 9-2: New research and translation campus in White City, West London.

50 spin-out companies, the Hub will support innovation on an unprecedented scale in London (Figure 9-2). It will provide high specification, multidisciplinary research space for scientists and engineers and state-of-the-heart incubator space for spin-out companies. Plans to develop the Research and Translation Hub began in 2013, and the site will eventually include leisure and retail facilities, a conference center, homes for College and Imperial NHS Trust key workers and for private sale, and a publicly accessible square (Imperial College West, 2014).

Alongside the award from HEFCE, one of the largest awarded by the UK RPIF, the new £150 million Research and Translation Hub will be funded by a £90 million contribution from investor Voreda, with the remainder funded by the College. Voreda Capital LLP is a London-based real estate fund that invests in property investment and development opportunities. The hub is expected to be completed in 2015.

With the new Imperial West campus and its clear translation agenda, Imperial College will be able to have companies grow a lot bigger and stay a lot longer because you have more space to play with. (Director of Imperial College bioincubator, June 2013)

9.2.2. Bioincubators in London

In total, there are 20 bioincubators in the UK. In Figure 9-3, you can find an overview of the most important bioincubators in the UK.

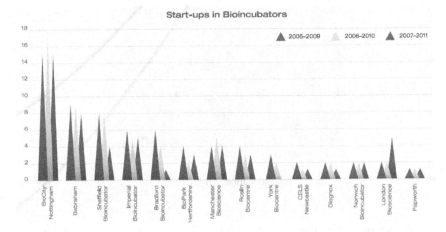

Figure 9-3: Overview of the most important bioincubators in the UK.

Figure 9-4: London BioScience Innovation Centre.

A total of 47% of UK bioscience companies formed in the last five years are based at a bioincubator site. Three of these are in central London, with a combined laboratory/office space of over 7,500 square metres: Imperial BioIncubator, London BioScience Innovation Centre and Queen Mary BioEnterprises Innovation Centre (Figure 9-4).

The London BioScience Innovation Centre (LBIC) encompasses 1,800 m² of office and laboratory space designed to support biomedical research and science. The creation of the London BioScience Innovation Centre in 2001 was a direct response to the evident shortage of accommodation for life sciences activity in Central London. As an established bioincubator the London Bioscience Innovation Centre has over 40 biotechnology and life science companies located on site.

A variety of different types of companies are based there, including international companies, small start-ups and more mature organizations. LBIC attracts biotechnology, life science product and diagnostic companies, contract research organizations, medical device companies and service providers. It is owned and operated by the Royal Veterinary College, one of the independent colleges of the University of London and a world-class veterinary and biomedical research institute. They are just a short walk from St Pancras International, making it easy to access markets both within the UK and mainland Europe. LBIC companies typically remain at the Centre for around three years, before graduating to a purpose built building or Science Park (London BioScience Innovation Centre, 2013).

The Queen Mary BioEnterprises (QMB) Innovation Centre (Figure 9-5) was officially opened in early 2011. This new facility covers 3,700 m² of state-of-the-art laboratory and office facilities. The Innovation Centre is located on the campus of Queen Mary, University of London in close proximity to its School of Medicine and Dentistry. The first tenants of the Innovation Centre signed up in June 2011. The tenants, Retroscreen Virology Ltd, were a Queen Mary spin out and one of London's largest biotechnology companies and Europe's leading anti-viral research organization. The QMB Innovation Centre is a wholly owned company of Queen Mary Innovation, the technology transfer company of Queen Mary University of London. It is the culmination of a four-year capital investment partnership between Queen Mary and the London Development Agency (Queen Mary incubator, 2013).

Figure 9-5: Queen Mary BioEnterprises (QMB) Innovation Centre.

9.3. Implementing best practices in your region

In this last part, we would like to focus on a few important guidelines for those who want to set up and manage a bioincubator. However, these best practices need to be implemented with great care and with a good knowledge of the university's structure and ecosystem. In other words, they need to be aligned with the university's structure and match the ecosystem of the university and the region. Only then could they also work in different regions and universities.

A first important best practice is the design of the incubator.

Some universities decide to use an existing building as an incubator. The design of it might not be ideal. The Imperial College bioincubator is designed in an 8 form, so that people have to pass other companies before they can exit/enter the incubator. So the design is crucial and facilitates the networking activities. It is all about community and to do that you need a well-designed space.

... So this also means that if you use existing buildings, you need to be really creative to turn the place into an exciting and dynamic incubator. (Director of Imperial College bioincubator, December 2012)

In line with this, the bioincubator can also facilitate networking activities by having lots of open areas where people can have discussions and (informal) meetings.

> There is always a high pressure to have as much letting space as possible. If you look around our incubator you will see a lot of open areas. These areas have white boards and everything that's needed to facilitate discussions and meetings. These are areas where companies/people can mix. This is also really important. (Hewson, 2012)

Third, the incubator puts together an exciting event agenda and organizes social events.

> Twice a month, we have seminar events. We invite people with different backgrounds and expertise, we invite legal people, marketing people,... whatever the topic of the seminar is. We also organize social events. We have a summer party and a Christmas party. Again it is important to create a community. (Director of Imperial College bioincubator, December 2012)

The bioincubator works closely with Imperial innovations to screen, assess, manage and invest in these companies.

> We have got a unique model in the UK. Imperial Innovations invested in more than 80% of our spin-outs so they are located very close to their investors. This also means that a certain due diligence is also executed before they come in. So there is a good chance that the company is going to do well. (Deputy Director of Imperial Innovations, June 2013)

The bioincubator should also be a place where academics meet business people, serial entrepreneurs and former CEOs.

> Imperial Innovations also recruits experienced CEOs to run these companies which means that the bioincubators is a place where you can meet people who set up several high-tech ventures. That mix works quite well. (Deputy Director of Imperial Innovations, June 2013)

Companies at the bioincubators can also use a lot of small services and benefit from economies of scale.

> Spin-outs do not have to invest in certain things such as a gym or health and safety training, ... Their people can go to talks and events at the

business school. They just use the services of the College and the incubator. They can access more for little. (Director of Imperial College bioincubator, December 2012)

Finally, every bioincubator needs its success stories in order to become more legitimate, attract high potential people and grow.

Until now Imperial sold already three companies which were located at the bioincubator, 38 companies already went through the incubator. Failures only three. These success stories are important for the growth of the incubator. (Director of Imperial College bioincubator, June 2013)

9.4. Best practices

- Design of the incubator that facilitates network activities and creates a community.
- A high-quality screening process.
- Organization of business and social events.
- Close collaboration with Imperial Innovations.
- Create economies of scale.

9.5. Future improvements

- The Imperial College bioincubator is full at the moment and has had to refuse a few spin-outs. Imperial West is being built to avoid spin-outs of Imperial having to go to other bioincubators in London.

9.6. References

Papers, reports, books, conference presentations and websites

Clarysse, B., Wright, M., Lockett, A., Van de Velde, E. & Vohora, A. (2005). Spinning out new ventures: a typology of incubation strategies from European research institutions. *Journal of Business Venturing*, 20(2), 183–216.

Hewson G. (2012). Guided tour in Imperial College's bioincubator, ETTbio site visit, Imperial College London.

Imperial Innovations (2013). Retrieved from: http://www.imperialinnovations.co.uk. Accessed on 12.01.2013.
Imperial College West (2014). Retrieved from: http://www3.imperial.ac.uk/imperialwest. Accessed on 06.06.2014.
London BioScience Innovation Park (2013). Retrieved from: http://www.lbic.com. Accessed on 20.10.2013.
London Medicine (2013). Retrieved from: http://www.londonmedicine.ac.uk. Accessed on 13.01.2013.
Queen Mary incubator (2013). Retrieved from: http://www.qmul.ac.uk. Accessed on 13.10.2013.

Case study interviews

- 8th December 2012, Director, Imperial College bioincubator, London.
- 10th June 2013, Director, Imperial College bioincubator, London.
- 21st June 2013, Deputy Director, Technology transfer team, Imperial Innovations, London.

10

CASE STUDY 10: Idea Lab — A Platform for Students to Develop New Ideas

Maarika Merirand and Siim Espenberg
(Tartu City Government, Estonia)

10.1. Setting the scene

Tartu has a population of around 100,000, making it the second biggest city in Estonia. It is the regional centre of southern Estonia and one of the country's two main academic and research centers.

Tartu accommodates the leading public universities — the University of Tartu and the Estonian University of Life Sciences — and other education and research centres. In total there are about 20,000 students studying in the higher education institutions of Tartu, and the major part of Estonian researchers and members of the academic teaching staff are located in Tartu. As the biotechnology industry in Estonia is strongly clustered around the universities, Tartu also hosts a range of biotechnology companies and related organizations.

The University of Tartu was founded in 1632 and is currently the only classical university in Estonia, covering a wide range of subjects. The university includes nine faculties, four colleges and a range of other sub-units, which have in total over 17,000 students, 1,400 PhD students and 3,800 employees, of whom 1,800 are research staff (University of Tartu, 2014).

Life sciences play an important role in the University of Tartu — the units most relevant to biotechnology are the Faculty of Medicine, Faculty of Science and Technology, the Estonian Genome Centre, and Tartu University Hospital. The main unit working with technology transfer is the Office of Research and Development, which consists of a Research Administration Unit, Research Project Unit, Intellectual Property Unit, Industrial Liaison Unit and Career Unit (University of Tartu, 2014).

In addition to developing academic excellence the university is increasingly focusing on improving technology transfer, encouraging entrepreneurship and strengthening cooperation with the companies. One of the recent initiatives addressing these issues is the Idea Lab.

10.2. Identifying best practices

10.2.1. *Idea Lab*

The Idea Lab was founded in 2011 and its aim is to offer students of all faculties an opportunity to implement their most outrageous ideas, try out new concepts and build prototypes in multidisciplinary teams (Idea Lab, 2014). The purpose of the Idea Lab is to raise the students' ability to implement their ideas in reality. It targets at the grassroots to promote an entrepreneurial mind-set. The students participating in the Idea Lab's idea development program are not pressured by expectations to succeed or to create companies, instead failure is considered as a natural part of the process (Manager, Idea Lab, June 2013).

The activities of the Idea Lab aim to develop the skills and general awareness of students of all levels. The vision of the Idea Lab is that in the future most of the students of the University of Tartu will participate in their programs at least once during their studies (Manager, Idea Lab, June 2013).

10.2.2. Earlier initiatives and role models

There has been a range of initiatives similar to the Idea Lab, developed within the University of Tartu. Some of these initiatives are listed below (Manager, Idea Lab, June 2013):

— A series of seminars led by the Tartu Information and Communication Technology Development Centre (TIKTAK) in the beginning of the 2000s, during which ideas were developed in multidisciplinary teams. One of the most remarkable outcomes of these seminars was the creation of Mobi Solutions, which today is the leading mobile development and services company in the region, with over 80 employees (Mobi Solutions, 2014).
— A project-based initiative Ettevõtluskodu (Home of Entrepreneurship) which was created in 2007 and is led by the Centre for Entrepreneurship of the Faculty of Economics and Business Administration. The aim of this initiative is to offer an innovative learning environment and to train people to draft projects and business plans, develop their management skills and encourage them to create jobs (University of Tartu, Centre for Entrepreneurship, 2014). The initiative has included ideas from different disciplines.

The initiative to create the Idea Lab came from University of Tartu's former Vice Rector for Research Mr. Kristjan Haller, who was inspired by the MIT Media Lab (USA) (Manager, Idea Lab, June 2013). The MIT Media Lab was founded in the 1980s and currently it involves 26 research groups and more than 350 projects. The aim of the MIT Media Lab is to "go beyond known boundaries and disciplines, encouraging the most unconventional mixing and matching of seemingly disparate research areas" (MIT Media Lab, 2014).

Another "role model" for the Idea Lab was Aalto University DesignFactory (Finland), which is a working environment that is designed for "flexible use with free interaction and prototyping made as easy as possible". It involves companies and researchers, but the main aim is to promote real hands-on learning for students and give them easy access to materials, equipment and tools (Aalto University Design Factory, 2014).

Upon developing the Idea Lab, the founding team gathered additional best practices from functioning incubators and similar initiatives in the Netherlands, Sweden and UK (Manager, Idea Lab, June 2013).

10.2.3. Foundation of the Idea Lab

The initiative to establish the Idea Lab arose from inside the University of Tartu, from the former Vice-Rector for Research Kristjan Haller after his visit to MIT in 2011. In addition to Mr. Haller, the following people contributed to the foundation of the Idea Lab (Manager, Idea Lab, June 2013):

— Mr. Indrek Ots, Head of the Office of Research and Development of the University of Tartu
— Mr. Erik Puura, Director of the Technology Institute of the University of Tartu
— Mr. Alvo Aabloo, Professor of polymer materials technology at the Faculty of Science and Technology of the University of Tartu, Member of the Scientific Board of Ahhaa Science Centre Foundation
— Mr. Mart Noorma, Vice-Dean of the Faculty of Science and Technology of the University of Tartu, Member of the Board of Ahhaa Science Centre Foundation
— Mr. Kalev Tarkpea, Director of the Institute of Experimental Physics and Technology at the University of Tartu.

Upon its founding the aim of the Idea Lab was defined as the following:

To create an environment developing youth's creative potential in the University of Tartu. The Idea Lab is an experimental lab for innovative ideas, incubating the students' most outrageous ideas and identifying ones that are implementable. The Idea Lab will bring together the most active students from all fields of the university and as wide a range of knowledge as possible. The Idea Lab gives a specific outcome to youth's creativity and enthusiasm through implementing some original and interesting idea by a spontaneously-forming interdisciplinary team. (Idea Lab, 2014)

Currently the Idea Lab is operating within the framework of the Office for Research and Development of the University of Tartu and is mainly financed from the central budget of the university, and employs three people. The manager of the Idea Lab is Mr Kalev Kaarna (Idea Lab, 2014).

10.2.4. *The working principles of the Idea Lab*

The Idea Lab works in cycles that coincide with the study semester at the University of Tartu. All ideas are developed during an 11-week period, and the program cycle is called TRAMM11. The Idea Lab's program is open to students of all levels. Students can submit their ideas for development via a web form. The ideas can be from any field, and in any stage of development, including:

— there is an identified problem, but no solution
— there are identified solutions, but it has not been defined which market segments they would offer value to
— the problem and solutions have been identified, but no implementation plan has been developed
— the problem and best solutions have been identified, but a prototype is necessary.

People who are interested in participating in the development of the ideas can also sign up via Idea Lab's website. If an idea gathers at least four interested students the Idea Lab will incorporate it into their program (Manager, Idea Lab, June 2013).

The Idea Lab's short-term projects increase necessary competences for doing product development with companies and with people from other disciplines. The 11-week process enhances necessary skills for bringing a new product, service or solution to market.

In some cases a project may last longer than 11 weeks and a team may continue to work on the project after the end of the program (Manager, Idea Lab, June 2013).

The Idea Lab enables teams to try out and use modern principles and tools for product development and project management (e.g. LEAN development, prototypes from paper, teams without managers, task management software) (Idea Lab, 2014). Equipped with such experience, university students are much better partners and employees for companies in the future. They've developed through this process a pro-active mind-set, time management, team building and management skills, all of which are valuable to every scientist, employee or entrepreneur.

The Idea Lab provides teams with support from academic mentors from fields specific to their ideas and also with support from entrepreneurs as well as access to resources (by providing, for example, use of laboratory space). At the end of each semester monetary prizes are awarded to the best teams, plus special prizes are given (Idea Lab, 2014).

Companies can use Idea Lab to source student teams to solve their problems or challenges. A contribution of 200 euros minimum is required from participating companies. However, upon initiating cooperation with the Idea Lab, a rewarding scale for participating students is established, according to the results.

The Idea Lab also involves the teaching staff of the university — for example, they can participate as field-specific mentors (which gives them access to students as potential future members of their research teams). In addition the academic staff is encouraged to include the opportunities provided by the Idea Lab as a practical exercise for the subjects they are teaching.

The Idea Lab has also started collaboration with incubators and universities in the USA, Great Britain, Finland and Latvia in order to give students multinational product development experience through virtual teams and 3–7 day training camps (Idea Lab, 2014).

In addition to the 11-week program, the Idea Lab has started another 11-month program called Student Company, which is more focused on building and enhancing the business skills and entrepreneurial mind-set of university students. The program supports students, who would like to establish (or at least try to run for a while) their own company. The students get training, personal coaching and mentoring from the Idea Lab. The process ends with a real or fictive closing of operations of the firm. In this way the students learn not only the starting and running of the company, but also closing of the firm. The aim of the program is to give to the students a real start-up experience (Idea Lab, 2014).

The Idea Lab also conducts weekly seminars about topics related to entrepreneurship and product development, which are open to all interested students (Manager, Idea Lab, June 2013).

The staff of the Idea Lab currently consists of three people — the manager, the mentors' supervisor and the project manager. The Idea Lab is predominantly financed from the central budget of the University of Tartu. In addition, a two-year project to set up a prototyping centre is financed by the Enterprise Estonia, the Estonian development and investment agency (Manager, Idea Lab, June 2013).

10.2.5. *Outcomes*

As the Idea Lab focuses more on increasing skills and awareness than pressuring participating students to achieve results and to establish start-ups, it is difficult to assess the tangible results. However, so far (by the first half of 2014) over 600 people have participated in the events organised by the Idea Lab. Over 50 student teams have been established in the framework of the Idea Lab, and around 20 teams have completed the 11-week cycle (Manager, Idea Lab, June 2013).

Since the foundation of the Idea Lab, the student teams have developed numerous physical and digital prototypes, several service concepts, and conducted experiments. The teams have worked on wide range of ideas such as developing new games for brain rehabilitation, using neural network algorithms for helping customers to make buying decisions, and using model copters for offering services to power line companies.

The work done by the teams within the Idea Lab has also been recognized outside the organization — some of the achievements include:

— reaching the final of "Ajujaht" (the biggest Estonian business plan competition, which takes place annually and the finals of which are broadcasted on national television)
— third place at the MIT Global Startup Workshop's Elevator Pitch Contest (MIT Global Startup Workshop, 2014)
— scholarship to a weeklong entrepreneurship camp in the Netherlands.

With regards to life sciences, Idea Lab is collaborating closely with "Entrepreneurship in Biotechnology", a course to ready master and PhD students within the Department of Science and Technology of the University of Tartu. The aim of the course is to give an overview of business opportunities in biotechnology, and it includes significant input from biotechnology entrepreneurs (Course "Entrepreneurship in Biotechnology", 2014).

One of the biotech ideas developed within the Idea Lab, GreenBead, was one of the top 25 business ideas of the business plan competition "Ajujaht" in 2013 (Ajujaht, 2014). The idea behind GreenBead is to create a reusable bioactive washing powder by tying the bioactive components to microcarriers and thus improve their bioactive functions.

Another more indirect effect to life sciences of the Idea Lab is the creation of the biotechnology entrepreneurship club. Currently there are also several life science projects in the Idea Lab's pipeline waiting for the teams to be assembled.

10.3. Implementing best practices in your region

The initiative to found the Idea Lab came from high-level inside the University of Tartu and different stakeholders from the university were involved in the creation of the Idea Lab — Rector's Office, research staff, faculties and the Office for Research and Development. This ensured that the working principles of the Idea Lab are aligned with the general priorities of the university. The Technology Transfer Office is also participating in the ongoing development of the Idea Lab (Manager, Idea Lab, June 2013).

The regional policy makers have not been directly involved with the creation and running of the Idea Lab, however, there is constant communication and cooperation by exchanging information, promoting and organizing events. For example, in 2012 and 2013 the Idea Lab has had an opportunity to hold seminars (targeting companies looking for solutions, etc.) within the Entrepreneurship Week organized by the City of Tartu (Entrepreneurship Week 2014).

10.4. Best practices

- The Idea Lab is a platform that offers students of all faculties an opportunity to implement their most extreme ideas, try out new concepts and build prototypes in multidisciplinary teams.
- The Idea Lab helps to promote the entrepreneurial mindset of students and also use their skills, ideas and knowledge to solve societal problems and needs.
- Problems, challenges, and ideas for solutions to be tackled in the Idea Lab come from companies, students, faculties and citizens.
- The students participating in the Idea Lab's idea development program are not pressured by expectations to succeed or to create companies, instead failure is considered as a natural part of the process.
- The Idea Lab also involves the teaching and academic staff of the university — for example, they can participate as field-specific mentors and use the Idea Lab as an environment where their students can put their theoretical knowledge into practice.
- The standard program of the Idea Lab lasts for 11 weeks and these short-term projects increase necessary competences for doing product development with companies and with people from other disciplines while trying out and using modern principles and tools for product development and project management.
- The Idea Lab also conducts weekly seminars, which are open to all interested students and has initiated an 11-month Student Entrepreneurship Program, which focused on building and enhancing the business skills and entrepreneurial mindset of

university students. The program supports students, who would like to establish or at least try to run for a while their own company.
- The Idea Lab has also a prototype center with different machines and other equipment that the students use to realize their ideas.
- The vision of the Idea Lab is that the students of the University of Tartu will participate in their program at least once during their studies.

10.5. Further opportunities

- The main problem of the Idea Lab is low awareness among students. Despite of having over 700 people on the mailing list, the rate of participation in the program could be increased. To address this the team of the Idea Lab are currently working on developing new communication channels to involve more students.
- Similarly, the involvement of companies in the work of the Idea Lab could be increased. The involvement of entrepreneurs creates more opportunities for participating students and provides additional financial resources for the work of the Idea Lab. Due to the uncertainty of results achieved by student teams, companies are hesitant to trust tasks to the Idea Lab. However, a number of companies have discovered that the student teams operating within the Idea Lab program can complement their own development activities and therefore contribute to innovation.
- Due to the specifics of life sciences, an 11-week program might not provide adequate time for teams to develop solutions in biotechnology. With the number of ideas developed increasing, and new biotechnology projects being included in the program, the Idea Lab might consider creating more field-specific programs in the future.

10.6. References

Papers, reports, books, conference presentations and websites

Aalto University Design Factory (2014). Retrieved from: http://www.aaltodesignfactory.fi. Accessed on 12.05.2014.

Ajujaht — Estonian Business Plan Competition (2014). Retrieved from: http://www.ajujaht.ee/. Accessed on 14.05.2014.

Course "Entrepreneurship in Biotechnology" (2014). Retrieved from: http://www.biotech.ebc.ee/Entrepreneurship_in_biotechnology.htm. Accessed on 14.05.2014.

Entrepreneurship Week (2014). Retrieved from: http://business.tartu.ee/. Accessed on 14.05.2014.

Idea Lab (2014). Retrieved from: http://ideelab.wordpress.com/idea-lab-in-english/. Accessed on 12.05.2014.

MIT Global Startup Workshop (2014). Retrieved from: http://www.mitgsw.org/. Accessed on 14.05.2014.

MIT Media Lab (2014). Retrieved from: http://www.media.mit.edu/. Accessed on 12.05.2014.

Mobi Solutions (2014). Retrieved from: http://www.mobisolutions.com/. Accessed on 28.05.2014.

University of Tartu (2014). Retrieved from: http://www.ut.ee/en. Accessed on 28.05.2014.

University of Tartu, Centre for Entrepreneurship (2014). Retrieved from: http://www.evk.ut.ee/. Accessed on 28.05.2014.

Case study interviews

03.06.2013, Interview with Kalev Kaarna, Manager, Idea Lab, Tartu (Estonia).

SECTION 4: EDUCATION

Introduction

Education of faculty members, postdoctorates and graduate students in technology transfer and entrepreneurship are important to enhance the effectiveness of technology transfer (Siegel & Phan, 2005). The Global Entrepreneurship Monitor (GEM) introduced a "value chain of entrepreneurship" (Bosma *et al.*, 2008). This value chain describes the process an entrepreneur goes through when starting a new company. The framework identifies several drivers for national entrepreneurial capacity and it is recognized that educational programs can influence these factors. Siegel & Phan (2005) suggest the development of a "Technological Entrepreneurship Curriculum" for technology transfer stakeholders (academic entrepreneur, TTO officer, incubator manager and small firm licensee). This curriculum should include elements such as entrepreneurship courses, technology familiarization, internships, idea labs, business plan competitions and a venture forum. The research institutions and universities play a crucial role in implementing such a curriculum at their organizations. In this section, we focus on best practices with regard to educational programs and initiatives that stimulate and support technology transfer and entrepreneurship.

Klofsten (2000) describes three basic activities that should be performed at universities in order to stimulate entrepreneurship: (1) activities that create and maintain an entrepreneurial culture;

(2) courses in entrepreneurship should be offered; and (3) specific training programs and coaching for potential entrepreneurs should be available.

When developing courses in entrepreneurship education, the main objective should be to increase the entrepreneurial qualification of students and researchers interested in technology transfer or starting a new venture. Jones & English (2004) explain that education should focus on skills that can be taught and characteristics that can be created, which are important in order to start a new company or commercialize intellectual property. To stimulate technology transfer and entrepreneurship, it is the aim of academic staff to identify and motivate the potential entrepreneurs and researchers interested in technology transfer.

Honig (2004) emphasizes the high impact of self-generated experiences or tacit knowledge for successful entrepreneurs. Braukmann & Schneider (2007) highlight that learning by doing and reflecting on the lessons learned should be connected and integrated in an entrepreneurship education program. Therefore, these authors propose educational concepts that allow students and researchers to try in practice and learn from their own mistakes. By doing so students and researchers are prepared for challenging business situations in the future.

In what follows, we discuss the best practices in education in technology transfer and entrepreneurship. Case Study 11 (VUB) describes how the three basic activities: (1) activities that create and maintain an entrepreneurial culture; (2) courses in entrepreneurship; and (3) training and coaching; are organized at the Vrije Universiteit Brussel in order to stimulate entrepreneurship and technology transfer. VUB has a technology transfer unit and deploys various initiatives that foster the awareness of technology transfer and create a supportive environment for academic entrepreneurship. A dedicated project team was established to develop a technology entrepreneurship education program for master's students and young scientists. Entrepreneurial expertise and experience is present within this project team. Next to this, dedicated people at the TTO are there to aid in the creation of spin-off companies as well as for providing professional support for industrial valorization, contract research and negotiations.

Case Study 13 describes the recently founded TTO "BioTech-IP" at the International Institute of Molecular and Cell Biology (IIMCB) and gives an example of a comprehensive educational program. The TTO offered courses focused on intellectual property protection, entrepreneurship, technology transfer as well as on soft skills. Case Studies 11 and 13 both describe best practice educational programs at an academic organization. However, Case Study 13 has a much shorter history as it was only started on 2010. In this case, the focus is more on courses. In comparison, Case Study 11 also highlights the entrepreneurial culture as best practice at VUB, which takes much longer to evolve.

Case Study 12 introduces a successful program (BioEmprenedor-XXI — Catalonia) for starting up and growing companies in the life sciences arena. BioEmprenedorXXI is a program designed to support and train entrepreneurial teams during the creation of their life sciences company, more specifically in the fields of biotechnology, biomedicine and agrifood. The program's mission is to promote the creation of high quality, innovative companies that are the future in the life sciences sector in Barcelona and Catalonia. The impact of the BioEmprenedorXXI program on the business fabric in the life sciences sector has been significant. One of the most significant, but also most debated, indicators for evaluating the results of any business start-up program is the number of companies created. In this sense, the program has had a high impact, as 51 of the participating projects (53% of the total) have ended up creating companies as of December 2013.

To summarize, this section shows best practices of comprehensive educational programs at academic organizations that stimulate entrepreneurship and technology transfer. It is important that a stimulating culture is created, courses for entrepreneurship and technology transfer are offered as well as individual coaching and training is provided. The section also introduces a successful training and coaching program from the region of Catalonia, Spain that supports and trains entrepreneurial teams during the creation of their life sciences company. This best practice shows the importance of a "learning by doing" environment.

References

Bosma, N., Jones, K., Autio, E. & Levie, J. (2008). Global Entrepreneurship Monitor; 2007 Executive Report. *Babson College, London Business School, and Global Entrepreneurship Research Consortium (GERA)*.

Braukmann, U. & Schneider, D. (2007). Die Entwicklung der Persönlichkeit des Unternehmers aus wirtschaftspädagogischer Perspektive. *Entwicklung unternehmerischer Kompetenz in der Berufsbildung–Hintergründe, Ziele und Prozesse berufspädagogischen Handelns, Bielefeld*, 93–121.

Honig, B. (2004). Entrepreneurship education: toward a model of contingency-based business planning. *Academy of Management Learning & Education*, 3(3), 258–273.

Jones, C. & English, J. (2004). A contemporary approach to entrepreneurship education. *Education+ Training*, 46(8/9), 416–423.

Klofsten, M. (2000). Training entrepreneurship at universities: a Swedish case. *Journal of European Industrial Training*, 24(6), 337–344.

Siegel, D.S. & Phan, P.H. (2005). Analyzing the effectiveness of university technology transfer: implications for entrepreneurship education. *Advances in the Study of Entrepreneurship, Innovation & Economic Growth*, 16, 1–38.

11

CASE STUDY 11: Entrepreneurship and Technology Transfer Education at the Vrije Universiteit Brussel

Tom Guldemont, Thomas Crispeels, Ilse Scheerlinck and Marc Goldchstein
(*Vrije Universiteit Brussel, Vesalius College, Belgium*)

11.1. Setting the scene

In this case study, we wish to highlight the technology transfer supporting activities that are organized at the Vrije Universiteit Brussel. In order to provide the reader with a comprehensive overview of these activities, we focus our discussion on the entrepreneurship education program set up by the team "Technology Entrepreneurship" (TE) and the communication activities of the Technology Transfer Interface (TTI) at the Vrije Universiteit Brussel.

11.1.1. *Flanders*

The region of Flanders is one of the three regions of Belgium. Situated to the north of Brussels, the capital city of both Belgium and Flanders, it is home to Belgium's Dutch-speaking community. Its approximately 6.3 million inhabitants live at the crossroads of the Netherlands, Germany, France and the United Kingdom, on a surface of 13,500 km² (2/3 of Silicon Valley). In addition, its capital Brussels is home to the European decision-making centers. Besides tax incentives, these are two crucial factors for a company when making an investment decision and choosing a location in Europe.

Flanders is a successful life sciences and biotechnology region with many biotech companies located within a small geographical area. Local and international investors are providing the capital for their continued growth. All competences required for bringing new and innovative products to the market are locally available. In Flanders, a life sciences investor will find a unique interplay between business, universities, research centers and hospitals. Good interaction and extensive networking between the various public and private life sciences players provides a dynamic environment, rich in innovation and knowledge sharing, in which new companies are constantly being added to a fast-growing life sciences cluster.

Belgium has four large, export-oriented clusters: chemicals, biopharmaceuticals, plastics and jewelry. Belgium is the world's second largest exporter of biopharmaceutical products. Strong factor conditions include a dense network of 167 hospitals and high-quality research institutions and universities (as measured by citations). This supports one of the highest pharmaceutical R&D re-investment rates in Europe. Belgium's favorable business context includes the fastest approval process in Europe for clinical trials (Phase I in less than two weeks). As a result, Belgium is the world's number one location for clinical trials per capita. The cluster also benefits from strong supporting industries such as biotech, chemicals and logistics, which complement industry activities in R&D, manufacturing and distribution. Flanders houses the biggest R&D hub for plant biotech — headed by the Flanders Institute for Biotechnology (VIB) — and emerging industrial biotech, with one of the largest integrated bio-energy production complexes in Europe — Bio Base Europe (Flandersbio, s.d.).

11.1.2. *Vrije Universiteit Brussel*

The Vrije Universiteit Brussel (VUB) is the coordinating university of the University Association Brussels (UAB), which consists of a strategic partnership between the VUB, the Brussels University Hospital (UZ Brussel) and the Erasmus University College (Erasmushogeschool Brussel, EhB) as key partners.

VUB is one of five Flemish universities and employs approximately 2,900 people, including approximately 1,500 FTE R&D staff, and has around 11,000 students. VUB consists of eight faculties, of which two are related to life sciences and biotechnology: the faculty of Science and Bio-engineering Sciences and the faculty of Medicine and Pharmacy. Furthermore, VUB is connected to the Flanders Interuniversity Institute of Biotechnology (VIB).

The university has two campuses. The main campus in Etterbeek houses seven faculties, while the medical campus in Jette is home to the faculty of Medicine and Pharmacy and to UZ Brussel, which is the University Hospital. The UZ has 721 beds, 28,200 hospitalized patients and 3,400 employees (Vrije Univeristeit Brussel, 2013).

11.1.3. *VUB TTO*

The VUB Technology Transfer Interface (TTI) started to grow out of the R&D department in the beginning of the 1990s, with the first steps in application-oriented research. The VUB Board of Directors approved the first valorization regulations in 1998.

It is not a coincidence that the Flemish government, by decree on 19 December 1998, started to provide subsidies to universities in order to establish Interface Offices at these institutions. These offices are charged by the Flemish government to implement interface activities in order to promote:

- cooperation between Flemish universities and industry
- the economic valorization of university research
- the creation of spin-off companies by universities.

The Interface subsidy, for a total of 2.8 million euros (2012), is currently allocated to the university associations according to the Interface allocation key. The Interface allocation key will be revised by the Flemish government in 2014, at the end of the current cycle.

Nowadays, the TTI is part of the VUB's R&D Department and consists of a multidisciplinary team of experts on technology transfer, business consultancy, contract negotiation, scientific funding, legal and IP issues, event organization and communication. They provide follow-up and advice to researchers at every phase of collaboration with third parties.

The Commission for Knowledge, Innovation and Technology Transfer (KITT) governs the Technology Transfer Interface. Comprising experts from both academic and the corporate world, the KITT Commission's mission is to develop a sustainable valorization policy at the university and to support individual technology transfer dossiers based on interdisciplinary knowledge and daily life experiences. As such, the TTI is not only a researcher's gateway to an industrial contact, but also a personal contact point for assistance and more information on many aspects of technology transfer. In that way, the TTI is responsible for the following activities:

- Finding the right funding for an applied research project.
- Management of the Industrial Research Fund (see Case Study 6).
- Protection and publication of knowledge and technology offers.
- IP strategy and business development of the IP-portfolio.
- Professional support for industrial valorization, contract research and negotiations.
- (Co)-Management of the university start-up fund QBIC.
- Business Plan coaching: the researcher should come up with a first proposal, while TTI makes some critical reflections to improve the plan.
- Finding investors: TTI may give advice about appropriate investors. It has good contacts with the QBIC fund, which has the right of first review on VUB spin-off dossiers, and has a broad network of venture capital contacts.
- The creation of spin-off companies.
- Providing information about incubators (IICB and ICAB) and finding office space.
- The organization of events focused on knowledge exchange and education.

- (Legal) Advice concerning contract research and administration.
- Creating a spirit of entrepreneurship.
- Managing an Industrial Network: CROSSTALKS.

In many of these activities the TTI plays the role of navigator and catalyst. The TT team shows the right way for the entrepreneur, but he or she is responsible for executing all steps required.

The VUB TTI is located at the university headquarters on the Etterbeek campus. It is a centralized office, serving all eight faculties of the university, covering a variety of different technologies. It fulfills its function with a team of 14 employees, of which four or five people are TT officers offering business development and valorization services, and providing legal and IP advice. The other members of the TTI are responsible for, among others, fundraising, research contracts, marketing and communication, and general administration.

In terms of educational background, two officers have a biotechnology related master's degree, while one TT officer holds a PhD in life sciences. There is one officer with an MBA. One officer is a Registered Technology Transfer Professional (RTTP), while two officers have a law degree or experience in law. Four to five people have industry experience, yet none in the biotechnology industry. On average, the officers have eight years of experience in the university's tech transfer activities.

Since the TTI team has to service all faculties, certain activities are outsourced. VUB patent applications are written and executed by an external patent attorney. Since the TTI cannot be fully familiar with the variety of different technologies that are being developed at the university, it also relies on external consultants and experts in the field, which is often very expensive (VUB, 2013; Hugo Loosvelt, personal communication, 2013).

The success of the TTI is for a large part evaluated by the parameters of the IOF Fund (Industrial Research Fund, see IOF Case Study). The money of this Fund is distributed among the Flemish universities based on parameters such as revenues from industrial collaborations, the number of spin-offs, the number of patents issued and the number of licenses granted. The amount of TTO "profits" or the number of products on the market are not part of the evaluation procedure of the TTI.

11.1.4. *Technology entrepreneurship at VUB*

The project "Technology Entrepreneurship at VUB", established in 2007, tries to address the (European) innovation paradox. The team, headed by Marc Goldchstein, is responsible for building and teaching an educational program on technology entrepreneurship and innovation.

The educational program addresses master's students (such as business, (bio-) engineering, computer sciences and so on) and young researchers and professionals at and around the VUB. The offer includes introductory entrepreneurship courses (such as the Starters Seminars), technology sector-specific courses (among others in biotechnology) and a master class in (technology) entrepreneurship.

Unique to the project is the opportunity to acquire real life experience. Students get the opportunity to work on concrete business development projects that originate directly from the research labs of the university (Crispeels, 2010).

Most team members work part time for the educational project. The other part of their job is within partner technology research departments, where they perform business development tasks, turning their teaching experience into practice (and vice versa). Project leader Marc Goldchstein shares his time between the educational project and the Technology Transfer Interface, where he supports research groups in their valorization strategies and coaches (potential) spin-offs. The team also provides coaching services to start-ups located at the university's incubator.

11.2. Identifying best practices

11.2.1. *Technology transfer interface*

TTI supports entrepreneurs in many different ways, not in the least by informing them about the different steps and aspects of technology transfer, going from the beginning of research to valorization and financial return. Hereunder, we discuss some of their most successful initiatives.

11.2.1.1. Booklet

A lot of information is compiled in a well-elaborated booklet that is developed and published by the TTI. This booklet is called "Knowledge, innovation and technology transfer issues: Finding your way through the jungle", and can be downloaded at the TTI website (VUB TTI, 2012).

Various steps and aspects of the technology transfer process are covered and are elaborated upon in this booklet, providing a researcher with useful information regarding relevant topics such as the valorization process, the TTI team, IP, the relationship between patenting and publishing.

Most of the information can also be found on the dedicated website of TTI, which was completely renewed in 2011: http://vubtechtransfer.be/. The website has several sections. Some sections are usual suspects such as "News", "Success Stories", "Calendar", "Partners" and "About us", while others offer dedicated portals for "Researchers" and "Companies". The "For Researchers" section provides all information that researchers and entrepreneurs need on tech transfer at VUB. The "For Companies" section gives a documented overview of the VUB's industrial policy, collaborations, business facilities, technology offers, business incubators and spin-off portfolio.

In the "Downloads" section, various brochures can be downloaded concerning the following fields: TTI services and tech transfer in general, VUB knowledge centers, VUB technology offers, VUB spin-offs and VUB research clusters.

11.2.1.2. Awareness creation, education and events

To create more awareness within the university about entrepreneurship, start-ups and contract research with the industry, the Technology Transfer Interface has been organizing seminars for more than ten years now, such as Starter Seminars, Technology Days and Contract Seminars.

Together with the project "Technology Entrepreneurship at the VUB" of the department of Business Economics and the Solvay Business School, the TTI organizes the Starter Seminars. These seminars provide a short but intensive course on the fundamentals of entrepreneurship and business economics. In eleven three-hour sessions, a variety of topics is covered, such as understanding business ecosystems, formulating strategies, developing a business plan, legal aspects, finance, marketing, the complex issues surrounding patenting, etc. Students of this program will acquire the vocabulary and mindset needed to think intelligently about their potential venture and thus interact more efficiently with tech transfer offices, investors and other partners. The Starter Seminars, which take place in the evening, are especially addressed to students and researchers of the VUB, though they are open to anyone interested. Under certain conditions the course is included in PhD students' Doctoral Training Programs. A couple of years ago, these seminars had 20 subscriptions, while nowadays there are more than 100 attendees. Presentations of the 2012 edition can be downloaded from the following website: http://vubtechtransfer.be/partners/teaching-technology-entrepreneurship-at-the-vub/.

11.2.1.3. Technology days

Know-how and technology scouting form part of the TTI's daily functioning. As expertise acquired through academic research is not always transferred to the outside world, valuable technologies and ideas, though not always developed for that purpose, may get lost for innovative developments. The TTI aims to bridge the gap between research at the university and the industrial landscape by regularly organizing Technology Days for new collaborations to emerge: thematic networking events for academic and industrial researchers, practitioners, policy advisors, innovation managers and venture capitalists. The event offers a great opportunity to learn about the latest state-of-the-art research in the field and consists of short presentations followed by round table sessions, giving the opportunity to catch up with the latest developments, ask questions and interact with researchers.

11.2.1.4. Contract seminars

On a regular basis, the TTI organizes seminars on the legal and administrative aspects of contract research for the VUB community.

11.2.1.5. Posters, calendars and brochures

Every year, the TTI distributes calendars and posters throughout the university. Each month on the TTI calendar features another tech transfer success story that took place at VUB and which demonstrates that entrepreneurial researchers are not that rare and have successful predecessors. The poster campaigns are aimed at researchers and invite them to contact (all contact details are displayed on the posters) the TTI whenever they have a question regarding their research and its valorization.

TTI has two structural partners: "CROSSTALKS" and the project "Technology Entrepreneurship at the VUB". Collaboration with these partners enhances the TTI's contacts with industry on the one hand and entrepreneurial students on the other hand.

VUB CROSSTALKS is a unique kind of academic and corporate networking launched by Vrije Universiteit Brussel in 2003. CROSSTALKS, the university and industry network of the VUB, operates as a neutral platform with a bottom-up and interdisciplinary approach. Through thematic encounters — congresses, workshops, publications, campustalks and Pecha Kucha Nights — CROSSTALKS aims at creating an open and constructive exchange between all stakeholders in society, beyond institutional and societal borders. Academics as well as CEOs, creative entrepreneurs, politicians, artists, architects and non-profit organizations are engaging in the CROSSTALKS networking concept (Crosstalks, 2013).

11.2.2. Technology entrepreneurship at VUB

Over the last six years, the technology entrepreneurship team developed a series of educational programs aimed at preparing and motivating business and technology students for a career in technology entrepreneurship. The deliverables of the project were a series of courses and practice projects offered to a range of audiences: bachelor and master's students in business engineering, (bio-)engineering, (computer) sciences; but also to young research and industry professionals in domains as biotechnology, life sciences and photonics (Crispeels *et al.*, 2009).

11.2.2.1. Partners

During its start-up phase (2007–2010), the project was funded by the Flemish government and industrial partners, representing a variety of industrial sectors (Amgen, Bank Degroof, Bekaert, Ethias, Fundus, IBM, Participatiemaatschappij Vlaanderen, Sirris, Solvay, Tyco Electronics and Yakult). The project is now further sustained with the support of the university as it is part of the General Strategic Plan of the VUB.

In 2006, the Flemish government, driven by Minister Fientje Moerman (at that time Vice Minister-President and Flemish Minister for Economy, Enterprise, Science, Innovation and Foreign Trade) decided to fund the project within the framework of the "bridge projects economics-education" (Brugprojecten Economie-Onderwijs).

In order to bridge the gap between theory and practice, the Flemish government wanted to support partnerships between education and the business world, with the aim of encouraging entrepreneurship among students, from primary schools to universities. The project "Technological Entrepreneurship at the Vrije Universiteit Brussel" was ranked fourth of 31 projects by the selection committee.

Within the VUB a strategic partnership was initiated between various faculties and the institute's technology research departments that were identified to have a substantial "industrial research" focus:

- The Faculty of Economic, Social and Political Sciences and the Solvay Business School
- The Faculty of Engineering Sciences
- The Faculty of Sciences and Bio-Engineering Sciences
- The Technology Transfer Interface.

Thanks to these partnerships, it was possible to employ project members/business developers who divided their time between the project (to develop courses) and the corresponding technological research group (to act as business developer).

11.2.2.2. Target group

The primary target group of TE are master's students in civil engineering and bioengineering. During their studies, these technological engineers get familiarized with the topic of business and entrepreneurship, such as writing a business plan, attracting funding and developing a marketing strategy. Unique to the TE approach is the fact that these business-economic trainings are tailored to the student's specific technology sectors, such as biotechnology, photonics, microelectronics and ICT, nano-chemistry and -technology; see Table 11-1 for the Table of Contents of the course "Business Aspects of Biotechnology" (2012–2013) which was taught to master's students in bio-engineering and business engineering.

Table 11-1: Course outline, "Business Aspects of Biotechnology", 2012–2013. Own set-up.

Nr.	Session	Speaker
1	Introduction and Industry Profile	Thomas Crispeels
2	Biotechnology Clusters and Business Models	Thomas Crispeels
3	Biotechnology finance (introduction)	Thomas Crispeels
4	IP in Life Sciences	Liesbeth Weynants — Hoyng Monegier
5	The Regulatory Framework for Biotechnology	Marc Martens — Bird&Bird
6	Personalized Medicine Business Planning	Thomas Crispeels
7	Business Development in Life Sciences	Tim Van Hauwermeiren — arGEN-X
8	Financial Management in Life Sciences	Wim Ottevaere — Ablynx
9	Pricing and Reimbursement	Valery Fikkert — Amgen
10	Case Studies: Starting up a Life Sciences Company	Thomas Crispeels

Technological engineers are not the only target audience of this project. Also the business students of the VUB get trained in the domain of technology entrepreneurship. These students choose one or several technology sectors and get an education in technology and business economics referring to that sector (Crispeels, 2010). The research departments of the VUB active in these areas provide much of the sectorial expertise.

11.2.2.3. *Educational program*

Over the past seven years the technology entrepreneurship team developed a series of educational programs aimed at preparing and motivating business and technology students for a career in technology entrepreneurship. The deliverables of the project were a series of courses and practice projects offered to a range of audiences: bachelor and master's students in business engineering, (bio-) engineering, (computer) sciences; but also to young research and industry professionals in domains as biotechnology, life sciences and photonics. For technological engineers, depending on the study field, certain courses are compulsory, while others are electives. For the business engineers, the basic offering is compulsory (Table 11-2). These programs contain the following core elements:

- An educational program on technology entrepreneurship for an as wide as possible audience of business and sciences/engineering students and young professionals; for the business students the educational programs starts in first bachelor year.
- A focus on specific technology sectors, through the development of courses on the business aspects of these technologies. The chosen domains are: biotechnology and life sciences, innovation in materials, photonics and micro-electronics, and software. These courses are offered to both business and (bio-) engineering students.
- Providing business and technology students the opportunity to perform real-life business development tasks on real-life technology projects, emanating from the research labs of the university and project partners.

Table 11-2: Overview of the technology entrepreneurship educational program at the Vrije Universiteit Brussel. Own set-up.

	Business engineers	Technological engineers
	Basic offering	
Bachelor	Introduction to Technology Entrepreneurship	
Master	Introduction to a technology* Master Class Technology Entrepreneurship Business aspects of a technological sector* Optional Technological Business Development Project	Introduction to Business

*Students in business engineering have to choose for a certain technological sector.

In a first phase, both groups of students get an introduction in the "other domain": an introduction to technology and engineering for business students, and an introduction to business and economics (Finance, Marketing, Human Resources and so on) for technology engineers. This is followed by a technology entrepreneurship course where students get acquainted with the reality of entrepreneurship, business ecosystems, venture capital, intellectual property, etc. Afterwards, business engineers and technology engineers follow a course on the business aspects of the relevant technology sector together. Optionally, the students work on the development of a business plan in the context of a Technological Business Development Project. They do this in mixed, multidisciplinary teams in order to ensure a certain level of cross-fertilization. During the academic year, they work on this project while being guided by the technology entrepreneurship team.

This offering has been incorporated in the relevant educational programs at the Vrije Universiteit Brussel and is being offered as "intensive training" to targeted audiences of young professionals and researchers, in collaboration with partners such as VIB, IMEC and EU FP7 programs.

An example is "Entrepreneurship in Life Sciences". The Team organizes this training in collaboration with VIB. This four-day entrepreneurship class is targeted at PhD students or postdocs in biotechnology and life sciences. The first day provides an introduction to the life sciences industry, business models and strategy, and offers an overview of the important components of a business plan. The second day addresses the basics of financial management in life sciences and elaborates on how to fund a life sciences company throughout its life cycle. On the third and fourth day the participants learn more about setting up a life sciences company: managing IP rights and human resources, negotiating with investors and closing deals with customers. Experts in the field present these sessions.[4]

11.3. Implementing best practices in your region

11.3.1. *Implications for TTO*

- Attract a dedicated marketing/awareness officer who takes care of the educational offers, events and awareness campaigns.

11.3.2. *Implications for universities/research organizations*

- Provide financial resources to the TTO to enable it to attract a marketing and communication officer.
- Support cooperation between different research faculties to join forces and to integrate dedicated entrepreneurship courses into various educational tracks.
- Allow for/facilitate the integration of educational projects and project members without an academic background into a research group.
- A business school facilitates the workings of the technology entrepreneurship project. Not only in terms of having qualified lecturers in the neighborhood, but also to enable cross-fertilization between students with a different background.

11.3.3. *Implications for policy makers*

- Support/fund initiatives that bring together organizations from the academic and the industrial world with the purpose of building an entrepreneurship education program (see call "bridge projects economics-education").

---

[4] More info on the workshop: http://www.vib.be/en/training/research-training/Pages/VIB-VUB-Workshop-Entrepreneurship-in-Life-Sciences-.aspx.

- Provide the necessary subsidies to support Technology Transfer Offices in their marketing and awareness creation activities.
- Integrating serial entrepreneurs or business developers (= often lacking a PhD or academic output mindset) in an academic institution poses some challenging issues with regards to responsibilities and evaluation. These issues need to be addressed at a regional level.

11.4. Future opportunities

- Funding: The project "Technology Entrepreneurship at VUB" was depending on financial support from the Flemish Government and industrial partners in the start-up period 2007–2010. During those years it was possible to attract the required human resources. However, after that period the financial support came to an end. Currently the project receives considerable financial backing from the Faculty of Economics and its research groups, but in order to take new initiatives, new sources of external funding will need to be found.
- Profiles and team composition: except from project leader Marc Goldchstein and consultant Jacques Langhendries, only junior business engineers were attracted to the technology entrepreneurship project. This means that considerable efforts had to be made for these people to become acquainted with their core technology, which takes time. Either this time should be given to junior researchers/business developers, or more experienced personnel with sectorial expertise should be attracted.
- The technology entrepreneurship project requires the proximity of a business school. Not only to have a mix of students, but also to have experienced lecturers in business topics. Physical proximity of other faculties (engineering, sciences, etc.) has proven to be very conducive to the success of the project.

11.5. Best practices

- Dedicated people at TTI responsible for Marketing and Communication. This enables various initiatives to foster the awareness of technology transfer and to create a supportive environment for academic entrepreneurship.

- A dedicated project team was established to develop and entrepreneurship education program with a strong technology focus. Entrepreneurial expertise and experience is present in this project team.
- During their Master studies, (technology) students get acquainted with entrepreneurship and technology transfer.
- Cross-fertilization between students with a different background is stimulated and facilitated.
- Setting up a project like technology entrepreneurship requires partnerships within the university, which strengthens the internal network and contacts. Support from the university community is greatly appreciated.

11.6. References

Papers, reports, books, conference presentations and websites

Crispeels, T. (2010). Technologische Ondernemers of Ondernemende Technologen. *EWI Review*, 10, 12–13.

Crispeels, T., Uecke, O., Goldchstein, M. & Schefzcyk, M. (2009). Best practices for developing university bioentrepreneurship education programs, *Journal of Commercial Biotechnology*, 15(2), 136–150.

Crosstalks (2013). Retrieved from: http://crosstalks.vub.ac.be. Accessed on 20.1.2013.

Flandersbio (s.d.). 10 Reasons to set up a Life Sciences R&D entity in Flanders, Belgium. Retrieved from: http://flandersbio.be/files/Factsheet_FINAAL.pdf. Accessed on 8.6.2013.

TTI Vrije Universiteit Brussel (2012). Retrieved from: http://vubtechtransfer.be/. Accessed on 17.11.2012.

Vrije Universiteit Brussel (2012). Retrieved from: http://www.vub.ac.be. Accessed on 19.11.2012.

VUB TTI (2012). Knowledge, innovation and technology transfer issues: Finding your way through the jungle. Retrieved from: http://vubtechtransfer.be/medialibrary/Knowledge-innovation-technology%20transfer%20-%20issues-2012.pdf. Accessed on 21.11.2012.

Case study interviews

- Hugo Loosvelt, 2013, TTI Vrije Univeristeit Brussel.

Additional online resources

- http://www.vub.ac.be/infoover/onderwijs/technologis-chond-ernemen/.
- http://vubtechtransfer.be/partners/teaching-technology-entrepreneurship-at-the-vub/.

12

CASE STUDY 12: BioEmprenedorXXI: Guidance Program for Starting up and Growing Companies in the Life Sciences Arena

Carlos Lurigados
(Biocat, Spain)

12.1. Setting the scene

Over the past ten years, Catalonia has seen an unprecedented level of growth in its research capacities, putting it among the most dynamic and productive regions in Europe in terms of knowledge generation. With a population of 7.5 million (Instituto Nacional de Estadística, 2014), Catalonia accounts for 1.5% of Europe's population; it contributes to 1.69% of its GDP (200,323 billion euros), 2.98% of scientific production and boasts 3.48% of ERC grants (ERC grants database, 2011), two-fold the European average, only preceded by Switzerland, Israel, UK and Belgium. Furthermore, joint efforts of both the Catalan and Spanish governments have allowed for the construction of key scientific facilities such as the National Genome Analysis Center (CNAG), the Alba-CELLS Synchrotron and the Barcelona Supercomputer Mare Nostrum.

Specifically, the BioRegion of Catalonia has 512 companies (biotechnology, pharmaceutical, innovative medical technology and sector services firms), 54 research centers, 16 science parks, 17 hospitals with research programs in the biosciences and 11 universities. These hospitals, universities and centers have joint total of 573 research groups, which especially stand out for their work in the fields of oncology and neurodegenerative diseases, areas where the wealth of the Catalan biotech business sector is also concentrated (Biocat Report, 2013).

Catalonia has 21% of Spanish biotech industry, 45% of Spanish pharma industry and 50% of the med tech sector (Asebio Report, 2013). It has 56 bioscience research centers, some of which are international benchmarks in genomics (CRG), photonics (ICFO) or oncology (IRB). Alongside this scientific development, the business sector has experienced unstoppable growth since 2000 — at a rate of 10% to 15% per year — and is now made up of 512 companies, including biotechnology, pharmaceutical, medical technology and sector services firms. This business activity is concentrated (85% of the organizations in the sector) mainly around Barcelona, which is also home to most of the 20 science parks in the region. Although the majority of these companies (65%) are SMEs and despite the adverse economic conditions they have had to deal with over the past years, the comparative analysis Biocat carried out for the 2009 and 2011 fiscal years shows a positive evolution both in economic terms and in research capacities (Biocat Report, 2013)

Catalonia leads Spain in number of start-ups, as well as in the creation and manufacturing of innovative products in this sector, both in the public and private sectors. If we take into account the dynamic nature of the creation of new biotechnology companies, Catalonia is also well above the Spanish average with eleven new companies in 2010 alone and a steady increase ranged from 10% to 15% per year over the past five years (Biocat Report, 2013).

For some years now, the life sciences arena (biotechnology, biomedicine and the agrifood sector) has been in the spotlight for the country's economic and social stakeholders. The role it plays in the development of industry, the benefits of its activity, including the prevention and treatment of disease, and the volume of investment it can raise (among many other factors) make this sector a driving economic force in Catalonia and, thus, a strategic sector for the economy.

With this aim, "la Caixa", Biocat and Barcelona Activa, decided to promote a joint initiative called BioEmprenedorXXI to contribute to the development of the sector's productive fabric, transferring the results obtained through research to society by creating new companies.

12.2. Identifying best practices

BioEmprenedorXXI is a program designed to support and train entrepreneurial teams during the creation of their life sciences company, more specifically the fields of biotechnology, biomedicine and agrifood. The program's mission is to promote the creation of quality, innovative companies with a sense of the future in the life sciences sector in Barcelona and Catalonia. This would help to grow the business fabric in the biotechnology arena and to build its ecosystem. The specific aims are:

- To promote a benchmark program in Europe for starting up companies in the life sciences sector.
- To spread business opportunities in the life sciences sector in Catalonia.
- To spread entrepreneurial and business culture in the sector.
- To provide the necessary tools to enable research findings to be transferred to society and companies.
- Giving people with entrepreneurial spirit the right training in business management through guidance actions to analyze the feasibility of the business ideas, training to acquire entrepreneurial skills and behaviors, direct contact with stakeholders in the sector as well as facilitating access to capital.
- To attract and maintain talent from the rest of Spain and Europe so they carry out their entrepreneurial activity in the biotechnology arena in Barcelona and Catalonia.
- To generate successful benchmarks in entrepreneurial initiative in the biotechnology sector on an international level.
- To become a meeting point for stakeholders in the sector, complementing activities carried out by the current Catalan innovation system.

12.2.1. Organizational structure

Five organizational bodies govern the BioEmprenedorXXI program: Promoting Committee, Technical Committee, Scientific Direction, Technical Secretarial and Panel of Judges. Each body is charged with its own specific functions depending on the nature of its members to ensure each works correctly. The Promoting Committee is in charge of marking the strategic lines and actions. Advising on strategic sectorial contents is the task of the Scientific Direction and the Technical Committee. The Technical Secretariat plays an executive role and controls the functioning of the program. The Panel of Judges assesses and decides which projects are chosen as the winners of the BioEmprenedorXXI awards. This structure is the result of experience acquired over the five years of the program and addresses the need for an efficient, flexible and operative organization.

12.2.2. Program contents

BioEmprenedorXXI offers numerous activities geared towards the creation of new biotechnology companies. These activities can be classified into four groups, depending on the phase of the process and their aims: seizing opportunities, training entrepreneurs, creating a network of contacts and preparing for investment. Figure 12-1 shows the various phases of the program and the activities included in each one.

12.2.3. Phase I: Seizing opportunities

The program is geared towards university graduates, researchers, professors, PhD students and entrepreneurs with a business idea or a company in the early stage of their life cycle. All participants want to explore their entrepreneurial ability to create and start up their own business project, either as a spin-off of their institution or independently. In this first phase of the program, both sectorial associations and entities and universities, hospitals and research centers play a crucial role. They are in charge of detecting entrepreneurial projects

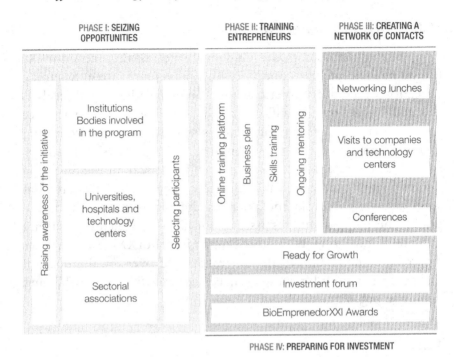

Figure 12-1:   Phases of BioEmprenedorXXI program.

that could potentially participate in BioEmprenedorXXI. In this sense, part of the success comes from this group's involvement in the program and their ability to detect new business projects. The selection process is carried out through personal interviews with those promoting the initiatives. The criteria to evaluate new projects focus mainly on the proposal's feasibility (covering a market need, business model, differentiation from competition, etc.) and on the promoting group's competences and skills (knowledge of the market, leadership abilities and commitment to the project, among others). In total a maximum of 20 business projects per year are selected.

12.2.4.   Phase II: Training entrepreneurs

One of the main aims of the program is for the entrepreneurs to assess the feasibility of their projects by creating a business plan. It must be noted that this feasibility doesn't only depend on the business idea but also on the promoting team. For this reason, the program devotes part of the training to making sure the team promoting the idea acquires the skills and capacities associated with entrepreneurship. The training content is made available to participants through two channels: online training using a telematics platform, and onsite training with face-to-face group sessions. This combination of formats (online and face-to-face) allows participants to combine their professional activity with the program. It must be noted that in addition to group training on how to create a business plan and in entrepreneurial skills, the program also includes personalized activities for each of the projects with individual mentoring sessions or tutorials.

12.2.4.1. *Online training: Creating a business plan*

The program provides with an online platform specializing in the bio sector. This platform is made up of eight modules with different themes (marketing, finance, legal issues, etc.) and participants have open access to all content and documents. Of these eight modules, four are required: marketing, legal and fiscal issues, strategy and financial analysis, as they are the most important in creating a business plan.

12.2.4.2. *Group face-to-face training: Creating a business plan and acquiring entrepreneurial skills*

The program includes a total of six face-to-face sessions. Each of the sessions has the same structure: in the morning participants look at various topics associated with business plans and strategy, then a networking lunch with a stakeholder from the sector, and in the afternoon workshops to acquire entrepreneurial skills. What follows is an outline of the sessions in the program. Each edition has had different topics, chosen based on the characteristics of the group: subsectors of activity, maturity of the projects, etc.

12.2.4.3. Individual face-to-face training (mentoring): Creating the business plan

BioEmprenedorXXI is designed to be an itinerary tailored to the needs of each business project. In this sense, each participant is assigned a mentor, who is in charge of guiding them through issues associated with creating their business plan. The mentor is an expert in the sector who guides the entrepreneurial team throughout the program and provides relevant information to help them create their business plans. They are in charge of analyzing the business ideas and their potential; supporting the entrepreneurs in detecting strengths and weaknesses; providing information on the market, competition, etc. The mentoring process is organized into four meetings with the entrepreneurial team after the mentor has reviewed and analyzed the documents submitted by each group. At these meetings mentors review the progress of the project, listen to the concerns of the entrepreneurial team, solve problems and propose improvements to the business plan. The mentors were added in the fifth edition of the program to address the requests of participants in previous editions. Previously the program offered only one mentoring session. The current mentoring program ensures guidance throughout the process and that mentors are more involved in the projects.

12.2.5. Phase III: Creating a network of contacts

One of the aims of BioEmprenedorXXI is to become a meeting point for entrepreneurs and stakeholders in the sector. Thus, the program incorporates a series of activities geared towards facilitating networking among stakeholders in the sector, like for example networking lunches and visits to technology parks and research centers.

12.2.5.1. Networking lunches

This is a joint lunch for participants in the program and one or more relevant stakeholders in the sector. The person invited may be of interest for various reasons: their knowledge of the sector and industry, their entrepreneurial experience, their international experience, etc. A total of six networking lunches are organized where entrepreneurs have the opportunity to exchange opinions, points of view and discuss other issues of interest with the guests.

12.2.5.2. Visits to technology parks and research centers

The aim of these visits is two-fold: to make participants aware of the reality of the business fabric and to facilitate the creation of networks of contacts. The entrepreneurs can thus get a first-hand look at how the most cutting-edge companies or research centers in each field operate and their strategies. The aim is for this activity to become something more than just a visit to the facilities and is thus carried out with the collaboration of a partner or director who, in sharing their knowledge and personal experience, brings value added to the visit. The idea is for this person to interact with the entrepreneurs, answer their questions, offer suggestions, explain current trends, etc.

12.2.6. Phase IV: Preparing for investment and start-up

The BioEmprenedorXXI program includes a series of activities designed to facilitate the search for funding so that participants can initiate their business activity.

12.2.6.1. Ready for growth seminar

Ready for Growth in the health sector is an investment-readiness seminar geared towards entrepreneurial projects and established companies.

This seminar aims to teach entrepreneurs and companies about the nature and structure of the process to obtain capital and the strategies to follow. Thus, participants have the opportunity to present their projects to the main investors in the sector in addition to learning important information on the funding process. The seminar especially emphasizes the different funding models available, the nature of different sources of funding (public capital, business angels, venture capital, etc.) and the relationship between investors and the companies they invest in. Ready for Growth normally includes participation from both public (ACC1Ó, ENISA, CDTI, etc.) and private (Caixa Capital Risc, Ysios Capital Partners, Inveready, etc.) institutions. It must be noted that the call for participants in this activity is open to the public (not just those participating in the program) and is geared towards projects that are sufficiently mature as to receive investment.

12.2.6.2. Healthcare Barcelona investment forum

Under the framework of the program, the Healthcare Barcelona Investment Forum is organized in collaboration with the Barcelona Medical Association, Keiretsu Forum, Barcelona Activa, IESE Business School and Biocat. These forums are characterized by their sectorial nature, as they are geared exclusively towards companies or business projects associated with biotechnology, medical devices, healthcare services and bioinformatics. As with Ready for Growth, this is an activity open to the general public, although the number of participating companies is normally restricted to a maximum of ten.

12.2.6.3. BioEmprenedorXXI award: Presentation of business plans

Each year, the program includes a session in which the business plans are presented to the Panel of Judges so they can see each of the participants and choose the winner and runners-up for the BioEmprenedorXXI Award. The session is divided into two parts. A first part in which all of the participants present their project to the Promoting Committee and Technical Committee. During the second part, the five best projects are presented to the Panel of Judges. The Promoting and Technical Committees select the five projects to be presented to the Panel of Judges, which selects the winners of the award. It must be noted that the role this body plays in this part of the program is key, contributing their sectorial and technical knowledge to evaluate and select the best projects.

12.2.7. Results of the program

The impact of the BioEmprenedorXXI program on the business fabric in the life sciences sector has been significant over its six years. In this time a total of 160 projects have been submitted, which is quite significant taking into account the specificity of the sector. Of all the applications received, 59% have participated in the program, meaning that BioEmprenedorXXI has worked with a total of 95 business projects.

One of the most significant indicators in evaluating the results of any business start-up program is the number of companies created (Table 12-1). In this sense, the program has had a high impact, as 51 of the participating projects (53% of the total) have ended up creating companies as of December 2013. The companies that have come out of the program have generated a total of 232 new jobs, which is significant taking into account the type of companies they are. If we take into account not only the companies that have already been formally established but also those in the process of doing so, the percentage goes from 53% to 63% (60 companies). This shows that one of the program's goals is being met: to facilitate the transfer of research findings to the business arena and society. Additionally, it must be noted that only 13 of the business projects have been abandoned. In analyzing why some of these initiatives didn't consolidate, the following reasons stood out: inability to associate the idea with a specific market, technical difficulties, and a lack of dedication from the team in developing the project. Finally, it must be noted that five of the projects, although they didn't lead to the creation of a company, are exploring alternative paths to take their idea to market, mainly through licensing their technology. Table 12-1 shows the characteristics of these companies and their impact.

Most of the projects focus on the medical technology and biotechnology arenas, with 29% and 24% of the total respectively. Additionally, 20% of total projects focus on pharmaceutical sector. Practically all of the other initiatives are distributed equally (between 2% and 10%) among the following areas of activity: food, bioinformatics, advanced therapies and services.

Table 12-1: Characteristics of companies created.

Companies Set Up	Jobs Created	Funds Raised
51	232	33 million euros
Patent Applications	Patents Registered	Companies with Products/ Services on the market
55	12	27
Companies with internationalization Plans		
32		

Regarding the business model, data shows that the mixed model, which combines a product and a technological platform, represent 36% of the total sample. Exactly the same percentage (36%) of companies are exclusively devoted to developing a product. With the final 28% of these companies focusing their business model on exploiting technology, a technological platform and/or a service.

In analyzing the investment received by companies once they had begun their activity, it can be seen that 45% of the companies established (23 companies) were able to raise funding. Being able to raise capital is a highly relevant indicator of both the quality of the program and the potential of the participating business projects. The total funding raised by participating companies is around 33 million euros.

12.3. Implementing best practices in your region

The public/private institutions promoting BioEmprendedorXXI are satisfied with the results achieved in six editions: 95 projects participating, of which 51 have started up innovative companies and nine of which are in the process of doing so.

An initiative like this is based on various factors that make it continue on successfully. Factors like the aims of all the partners being fully aligned, which is to say that they share a vision and mission that the only way to generate wealth in an area is by creating companies that contribute value to the market. Another factor is that there is a training program that has evolved over the years and that, today, can be said to give all participants the necessary tools to start up a business project, the necessary management skills and priority access to contacts and funding. With the desire to continue improving, the number of personalized mentoring sessions has been increased to four in order to allow for a longer guidance period and more participation from the mentor. At the same time, the number of experts participating in the lunches organized for participants and stakeholders in the sector has doubled. This activity is one of the program's great assets as it fosters debate in a relaxed setting, far from the formality of training sessions. And, last but not least, there was the opportunity to accompany 51 new projects through the process of taking their research to market, and that benefits society and creates innovation that helps make the world a better place.

12.4. Future opportunities

- Incorporate new activities and contents, and increase the level of internationalization by attracting entrepreneurial initiatives from other countries that want to establish and grow their companies in and from Catalonia.
- Develop the program to bring it closer to international acceleration and incubation programs, meaning programs supporting start-ups with a strong component of private promoters, in which entrepreneurs are guided over a period of two years.

12.5. Best practices

BioEmprenedorXXI is a successful program to support the creation of companies based on the following success factors:

- Aims, vision and mission of all the partners are fully aligned in order to generate wealth in an area by creating companies that contribute value to the market.
- A training program that gives all participants the necessary tools to start up a business project, the necessary management skills and priority access to contacts and funding.

- Accompanying new projects through the process of taking their research to market.

12.6. References

<u>Papers, reports, books, conference presentations and websites</u>

Asebio (2013. INFORME 2013 ASEBIO: Situación y tendencias del sector de la biotecnología en España.

Biocat (2013). 2013 Biotech Report: Catalan Life Sciences — Status and Analysis: Commited to Value Driven Growth. Barcelona.

ERC grants database (2011). Retrieved from: http://erc.europa.eu. Accessed on 08.02.2012.

Instituto Nacional de Estadística (2014). Retrieved from: www.ine.es. Accessed on 22.07.2014.

12.7. Acknowledgements

The entities that promote the BioEmprenedorXXI program would like to thank Almirall, the Barcelona Science Park, Reig Jofre, Grifols, the Catalan Foundation for Research and Innovation, Osborne & Clarke, and Vir Audit for their support as sponsors. Additionally, BioEmprenedorXXI wouldn't have been possible without the support, interest and participation of all the stakeholders in the innovation ecosystem: universities, research centers, hospitals, and companies, among others. We also thank the speakers, experts and members of the Panel of Judges and Technical Committee, and in general, on a personal level, everyone who has been involved in this initiative. And, finally, we especially want to thank all the bioentrepreneurs that have participated in this program and made it what it is.

BioEmprenedorXXI is an initiative of:

Promoted by

Sponsored by

13

CASE STUDY 13: Education for Scientists

Magdalena Powierża, Piotr Potepa and Hubert Ludwiczak
(International Institute of Molecular and Cell Biology, Poland)

13.1. Setting the scene

The Polish scientific and academic potential is concentrated in 460 universities and higher education institutes and in 200 research and development centers. Altogether they employ nearly 100,000 research and scientific staff and teach two million students. Research and scientific institutes may count on the financial support of the government, available EU funds and support of scientific institutions (Gwiazda et al., 2012).

The highly innovative activity of enterprises and fruitful use of knowledge and scientific research by the industrial sector are the key competitive factors, both on the national and regional levels. The scale of creating and absorbing innovations is highly unsatisfactory in Poland. The low involvement of business in financing research and development field shows that there is a lack of cooperation between industry and R&D. It also indicates a structural weakness of the R&D sector in Poland. Investments in young scientific personnel, realized by means of scientific doctoral scholarships, seem to be required to improve innovativeness and enhance competitiveness.

Ranked first among all scientific institutions in the field of biological sciences in Poland is the International Institute of Molecular and Cell Biology (IIMCB); located at the Ochota Biocentre in Warsaw. The Institute began its independent scientific activity on 1 January 1999. Currently there are about 140 researchers aboard, including 50 PhD students. IIMCB is made up of nine research groups, including a joint one with the Max Planck Institute of Molecular Cell Biology and Genetics in Dresden, Germany. The research focus at IIMCB is fundamental biomedicine. The Institute is financed in part from the national budget and in part from other sources (Ministry of Science and Higher Education, Foundation for Polish Science, National Science Center, National Center for Research and Development, Framework Programs of EU, Max Planck Society, Howard Hughes Medical Institute, European Molecular Biology Organization, National Institutes of Health, Wellcome Trust, European Science Foundation, etc.). About 80% of funds are competitive grant awards received by group leaders (IIMCB, 2014).

The highest quality of science achieved at IIMCB gave rise to the establishment of Bio&Technology Innovations Platform (BioTech-IP) — the Technology Transfer Office within the IIMCB structures (see also Case Study 3). BioTech-IP renders its services to all research institutes within the Ochota Biocentre Consortium. The main goal of BioTech-IP is to support commercialization of inventions and technologies developed by scientists in six public research institutes affiliated to Ochota Biocentre Consortium, as well as providing support (training and workshops) to scientists in all matters related to patenting, IP-rights and R&D project management. Another important task of BioTech-IP is the promotion of intellectual and infrastructural potential of Ochota Biocentre together with its commercial services. This provides an environment that promotes collaboration between researchers and enterprises in the field of biomedical and life sciences. BioTech-IP's core activity is to support research at all stages of technology implementation: from idea, invention, processing of intellectual property protection, to setting up a spin-off company and licensing the intellectual property. In particular, the BioTech-IP's activities focus on searching for and verifying research projects with a strong commercial potential and on new technology market potential assessment (BioTech-IP, 2014).

Since its establishment in 2010, BioTech-IP has been involved in helping scientists to understand the process of commercialization of inventions by organizing workshops and training sessions covering topics such as patenting, IP protection and project management. BioTech-IP helps scientists to secure funds for applied research and provides financial support to Ochota Biocentre PhD students carrying out research projects with a high commercial potential. In 2012 BioTech-IP started an internship program for PhD students and established scientists on the campus. The program aims to create highly skilled and industry-savvy scientists poised to enter diverse career paths in industry and academia. Another activity of the BioTech-IP initiated in 2012 are the networking events, which aim to build a bridge between those who develop new technologies and those who are able to bring the benefits of innovation to the market. Ochota Biocentre has great scientists, yet due to insufficient training in technology transfer and soft skills there exists a high demand for courses on these topics among Ochota Biocentre students and scientists. That is why BioTech-IP paid a lot of attention to their needs and offers specially adapted trainings. The following section describes various education programs and networking events for students and researchers of the Ochota Biocentre.

13.2. Identifying best practices

Based on a high demand for education in technology transfer, entrepreneurship and intellectual property rights Bio-Tech-IP participated in an EU project — Measure 8.2.1 Support to cooperation of scientific environment and enterprises (European Funds Portal, 2014). Within this project several best practice actions devoted to increase entrepreneurial skills among scientists have been developed and implemented for the Ochota Biocentre. These educational courses are described as follows.

13.2.1. Educational courses for scientists provided by TTO

As mentioned before, scientific staff at Ochota Biocentre needed training on technology transfer and IPR. BioTech-IP responded to such needs by organizing courses on various topics such as the following (BioTech-IP, 2014).

13.2.1.1. Course: "Workshop on scientific communication — interdisciplinary grants"

Communication is a very important part of research performed by scientists. It gives crucial competences for a successful career and helps understand basic communication strategies. After this course scientists should be able to find answers to such questions as:

- What information should be included in an abstract?
- How much text is acceptable on a poster?
- What is the best way to get a message across in an oral presentation?
- How should you present your ideas in a concise and communicative manner to non-scientific audience (e.g. the media or investors)?

13.2.1.2. Course: "Everything you don't know about patents"

This two-day training course is offered usually at the beginning of the academic year in order to give all new PhD students a chance to learn about the IPR basics. The workshop has very practical dimensions — participants search databases with respect to the area of science they work on in their projects and learn to read patent documents. Besides general information about patents, scientists receive personal help from patent attorneys. They can ask questions related directly to their projects and problems. It is a great opportunity to get free legal advice.

13.2.1.3. Course: "Research funding"

This information course gives a wide overview on possible funding sources in the coming months. Representatives of funding institutions present available programs with a detailed description of provided support.

13.2.1.4. Course: "Effective communication and self-presentation"

Small things play vital roles such as non-verbal communication through dress code, grooming, etiquette and body language. This innovative class teaches how to:

- create a positive first and lasting impression
- use techniques that make your communication straightforward and more efficient
- increase confidence, capability, credibility, productivity and individuality.

13.2.1.5. Course: "Team management"

Team management is an important skill for lab leaders. It helps achieve the effectiveness of groups of individuals within an organization. Participants can learn how to inspire a team, ensure team roles but also how to deal with possible conflicts.

13.2.1.6. Course: "Negotiations with business partners"

This course is designed to improve skills in all phases of negotiation. It helps in understanding general negotiation theory and buyer–seller relations. Participants get information about various agreements with business partners, focusing particularly on the most important issues.

13.2.1.7. Course: "Time management and personal effectiveness"

The aim of this course is to become more effective at work and learn how to better manage time by:

- identifying your own time management problems
- developing a more disciplined approach to work
- selecting priorities
- creating good habits in time management.

13.2.1.8. Course: "From the invention to the product — commercialization strategies in practice"

A training course led by a professional tech transfer manager who explains the process of commercialization and the role of scientists. It shows the relations between patenting and commercialization strategies and different types of commercialization (spin-off/licensing).

13.2.2. Science-to-business brunches

Science-to-business brunches is a new idea launched by BioTech-IP in 2013. The main purpose of this event is to tighten cooperation between researches, companies and investors. During the two-hour morning meeting, three or four scientists present their projects with possible industrial application. Each presentation is approximately ten minutes long. Scientists have to be very straight and market oriented. After the general presentation there is time for networking. Thanks to this event, researchers know what information should be involved to "sell" their projects and ideas. Representatives of companies are matched with presented R&D

projects. Overall, science-to-business brunches are an excellent opportunity to:

- meet potential research and business partners
- create new interdisciplinary relationships for future R&D projects
- network and finding contacts for future commercialization.

Brunches are mainly aimed at researchers, businesses and private equity investors (BioTech-IP, 2014).

13.2.3. Scholarships for PhD students

Scientific scholarships are awarded to PhD students who carry out applied research projects at Ochota Biocentre for a period of 12 to 36 months. Applications for Scholarships take place upon submission of an application form to the BioTech-IP. Candidate selection process is carried out by the Scholarship Committee, made up of scientific experts in the given fields, entrepreneurs and patent attorneys. The results achieved by the students are monitored on yearly basis. BioTech-IP assists the projects in acquiring patent protection and their further commercialization (BioTech-IP, 2014).

13.2.4. Paid internships for PhD students and scientists in enterprises

The purpose of internships is to strengthen the cooperation of graduate students of Biocentrum Ochota and enterprises to support the transfer of knowledge. During the internship a trainee performs tasks associated with research and development of the host company, which enables the exchange of knowledge and experience between science and business. Internships may be granted for a period of two months (not shorter than two weeks). Applications for Internships Program take place upon submission of an application form to BioTech-IP. Candidates should choose the host company, in which she/he intends to take internship with earlier consultation of term placement and research work schedule described in the individual plan for the internship (the company must be registered in Mazovia district). After the completion of the internship the trainees present a report and are paid an allowance (BioTech-IP, 2014).

13.3. Output/results/achievements

Before the program of training activities was implemented there was one spin-off company formed, one national patent granted and one international PCT application filed by IIMCB. Since the gradual implementation of the above-mentioned activities, which started in 2010, there is still one spin-off company created so far, but there are ongoing efforts to establish a new company from IIMCB. Nevertheless, the number of international PCT applications filed by IIMCB scientists has increased to seven in total.

Additionally, the training activities (courses and workshops on various topics related to IPR, PhD scholarships, internships and networking opportunities) offered to scientists by the BioTech-IP have within a fairly short time frame contributed to a better understanding of the commercialization process. The primary aims of these programs were:

- to concentrate all activities in one place — for over 1,500 scientists and more than 1,000 PhD students on the Ochota Campus with 14 scientific institutions within 1 square kilometer area
- to raise awareness on IPR and management issues helpful in the design of future grant applications or in establishing new high-tech ventures
- to promote collaboration with industrial partners
- to provide opportunities for "learning by doing" and "hands-on experience".

The results of all training activities as a number of participants (PhD students and scientists from Ochota Campus) undertaken since 2010 are summarized in Table 13-1.

Table 13-1: Results of training activities.

Activity	Year started	Total number of participants
Workshops, courses and trainings on patenting, IP protection, project management and soft skills	2010	856
PhD scholarships awarded (III rounds)	2012	30
Internships	2012	7
Science-business networking brunches	2013	108

Source: BioTech-IP.

13.4. Implementing best practices in your region

In this case study, we focus on best practices in high-tech entrepreneurship education for scientists with the aim to bring the researchers closer to understanding the market needs and rules of commercialization of scientific inventions. The decision to offer such a comprehensive package of courses and workshops coupled with the opportunity to gain hands-on experience arose from the need to train especially a young generation of scientists in all matters related to IPR and high-tech entrepreneurship which was accompanied by the availability of Structural Funds — the financial instruments of European Union (EU) regional policy. The other need was to offer centralized (concentrated on the Ochota Campus) set of training courses for researchers for which the venue in most cases was directly on the campus so they could benefit from the close proximity to their workplaces. However, the latter is not a strict condition as the courses may be available for other scientists from the region.

The best practices mentioned in this case study need to be implemented with great care taking into account that the main source of money for the courses, workshops, scholarships and internships organized by BioTech-IP comes from EU Structural Funds. European Union offers funding for various trainings. BioTech-IP is active in this field, regularly checking availability of possible grants and applying for them. It is important to know all the institutions responsible for implementing and organizing EU funds. Usually, the plan for upcoming calls for proposals is announced several months in advance.

Partners interested in this area are firstly obligated to check available sources supporting education for scientists on domestic labor market. However, the EU Structural Funds don't have to be the only source of funding. There are also opportunities to attract, for example, industry to work out joint training courses.

13.5. Future opportunities

- Attendance with closed group of people interested in courses.
- Training center for anyone not only from Ochota Biocentre.
- More focus towards "mini-MBA" intensive bioentrepreneurship courses.

13.6. Best practices

- Development of better relationships with the academic community.
- Comprehensive package of training activities.
- Good network of experienced trainers.

13.7. References

Papers, reports, books, conference presentations and websites

BioTech-IP (Bio & Technology Innovations Platform) (2014). Retrieved from: http://www.biotech-ip.pl/. Accessed on 27.06.2014.
European Funds Portal, Human Capital Programme (2014). Retrieved from: http://www.efs.gov.pl. Accessed on 27.06.2014.
Gwiazda M., Mazurek B., Zalewska A., Rebkowiec G. & Leśniewski Ł. (2012). Tax and Polish Information and Foreign Investment Agency, National Centre for Research and Development; "R&D market in Poland — support for research and development activity of enterprises".
IIMCB (International Institute of Molecular and Cell Biology) (2014). Retrieved from: http://www.iimcb.gov.pl/. Accessed on 27.06.2014.

Further reading

International Institute of Molecular and Cell Biology (2014). Annual Report 2013. Retrieved from: www.iimcb.gov.pl/tl_files/Reports/IIMCB_Report_2013.pdf. Accessed on 27.06.2014.

Ministry of Regional Development (2007). Operational Programme Human Capital 2007–2013. Retrieved from: www.efs.gov.pl/English/Documents/HCOP_EN_18January2008_final.pdf. Accessed on 27.06.2014.

Ministry of Regional Development (2009). Detailed description of the priorities of Operational Programme Human Capital 2007–2013. Retrieved from: www.efs.gov.pl/English/Documents/SZOP_wersjaangielska.pdf. Accessed on 27.06.2014.

SECTION 5: CLUSTERS

Introduction

The foundation of Genentech in 1976 near San Francisco marks the birth of the modern biotechnology industry. Today, the San Francisco bay area is still the beating heart of the industry and contains a high concentration of biotechnology firms, research centers, service providers and investors. More than thirty years later, biotechnology activity is concentrated in a relatively small number of regions such as San Diego (CA), Boston (MT) or Los Angeles — Long Beach (CA). In Europe, biotechnology clusters have emerged in, among others, UK, Germany, Belgium and Spain. Clusters do not only appear in the biotechnology setting. In a world characterized by decreasing trade barriers and transportation costs, operations are more easily centralized in few(er) locations and clustering has been observed to increase over time (Steinle & Schiele, 2002).

Interestingly, in the case of biotechnology the proximity of large anchor companies or customers such as big pharmaceutical or chemical conglomerates does not seem to be necessary for the formation of biotechnology clusters. Swann and Prevezer observe "clusters of small [biotechnology] companies developing near research centers rather than close to the established companies" (Swann & Prevezer, 1996, p. 1141). But then why do high technology activities such as the biotechnology industry cluster? As early as in 1919, Marshall summarized the advantages for a firm operating within a cluster as follows:

(1) the availability of common buyers and suppliers; (2) the formation of a specialized and skilled labor pool; and (3) the informal transfer of knowledge. Indeed, physical proximity results in easily available information and frequent interactions result in trust between parties. The latter allows for lower barriers to collaborate and in such a club-like atmosphere knowledge-exchange can be intense, resulting in higher levels of growth and performance (Maskell, 2001; Steinle & Schiele, 2002). Access to and integration of external knowledge is key to an organization's success, especially in a knowledge-intensive industry such as the biotechnology industry (Cohen & Levinthal, 1990; Grant & Baden-Fuller, 2004). The most important resource sought in external partners is knowledge (Zaheer et al., 2010).

Reaching a high (threshold) concentration of biotechnology research centers and companies thus seems to be conducive to develop regional biotechnology industry. Many local development agencies and governments thus aim to develop the biotechnology industry up to a point where it will gain a self-sustaining momentum, not in the least because clusters contribute disproportionally high to employment, GDP and exports.

Several determinants of biotechnology clustering or factors that make it possible for a biotechnology cluster emerge have been identified: (1) the presence of a strong scientific, technological and industrial base; (2) a favorable financial climate; (3) strong IP regulations; (4) a regulatory climate that does not restrict genetic experimentation; and (5) mechanisms that favor communication and transfer of knowledge between academia and the industry (Nelsen, 2005; Orsenigo, 2001).

In this section, we present four case studies of European regions that have put into place a number of measures/initiatives to support the development of regional biotechnology industry with the aim of creating a cluster. Open and intensive collaboration between the academic, industrial and governmental stakeholders runs as a red wire through these case studies. The case studies highlight how governments or regional development agencies have put into place effective communication and knowledge exchange platforms (Case Study 14 and Case Study 16), often in close collaboration with regional industry. We furthermore present a detailed story on the foundation of a successful biotechnology company (Ablynx) in Case Study 15. We observe once again that the role and vision of government agencies and politicians are crucial in the development of a biotechnology cluster and even in the foundation of a biotechnology company. This case also highlights the importance of communication and informal networks. The last case study (17) introduces the DRESDEN-concept and how a shared services and facilities approach can benefit regional stakeholders in their research activities. Interestingly, this concept, which is backed by the government, combines stakeholders from the scientific community with stakeholders from the cultural scene. All case studies highlight that supporting clustering activities is more than supporting formalized collaboration or scientific research: it is about building a community and supporting an environment in which valuable ideas and initiatives can germinate and grow.

References

Cohen, W.M. & Levinthal, D.A. (1990). Absorptive capacity: A new perspective on learning and innovation. *Administrative Science Quarterly*, 35(1), 128–152.
Grant, R.M. & Baden-Fuller, C. (2004). A knowledge accessing theory of strategic alliances. *Journal of Management Studies*, 41(1), 61–84.
Liebeskind, J.P. (1996). Knowledge, strategy, and the theory of the firm. *Strategic Management Journal*, 17(S2), 93–107.
Marshall A. (1919). *Industry and Trade*, London, Macmillan.
Maskell, P. (2001). Towards a knowledge-based theory of the geographical cluster. *Industrial and Corporate Change*, 10(4), 921–943.
Nelsen, L.L. (2005). The role of research institutions in the formation of the biotech cluster in Massachussetts: The MIT experience. *Journal of Commercial Biotechnology*, 11(4), 330–336.
Orsenigo, L. (2001). The (failed) development of a biotechnology cluster: The case of Lombardy. *Small Business Economics*, 17(1–2), 77–92.
Steinle, C. & Schiele, H. (2002). When do industries cluster?: A proposal on how to assess an industry's propensity to concentrate at a single region or nation. *Research Policy*, 31(6), 849–858.

Swann, P. & Prevezer, M. (1996). A comparison of the dynamics of industrial clustering in computing and biotechnology. *Research Policy*, 25(7), 1139–1157.

Zaheer, A., Gözübüyük, R. & Milanov, H. (2010). It's the connections: The network perspective in interorganizational research. *The Academy of Management Perspectives*, 24(1), 62–77.

14

CASE STUDY 14: The Biocat Model: Managing the Bioregion of Catalonia

Adela Farre and Carlos Lurigados
(*Biocat, Spain*)

14.1. Setting the scene

14.1.1. *Biocat*

Biocat is the organization that coordinates and promotes the life sciences sector in the BioRegion of Catalonia. Its mission is to support all the stakeholders in this area and their initiatives to boost research, knowledge transfer, business innovation and entrepreneurship in order to make biotechnology and biomedicine become driving forces for the country's economy and contribute to the wellbeing of society as a whole.

Created in 2006 at the behest of the Government of Catalonia and the Barcelona City Council, Biocat is a foundation that brings together representatives from all areas of the biomedicine and biotechnology sector: administrations, universities, research centers, companies and support bodies. Biocat, among others, compiles the requirements of society and the national and international markets in order to align the biocluster's strategy for the futures. To this end, Biocat works to detect the major strategic international trends for forecasting needs and opportunities for the sector before they even appear.

14.1.2. The BioRegion of Catalonia

The BioRegion is the life sciences cluster in Catalonia. It consists of biotechnology, pharmaceutical, medical technology and service companies (512) in addition to universities (11), research hospitals (17) and research centers (54). It also includes knowledge transfer support and innovation structures and networks. Service companies and suppliers represent a significant percentage of the life sciences firms active in the region (23%). The core life sciences companies (biotech, techmed and pharma) represent more than the 55% of the total of companies of the cluster (Biocat Report, 2013).

Barcelona, the capital of Catalonia, is the main biotechnology hub in Spain: 21% of Spanish biotech companies are located in Catalonia, most of them in Barcelona. During the last five years (Asebio Report, 2013), the rate of new biotech company creation in Catalonia has ranged from 10% to 15%. This rate is higher than the European average.

The BioRegion of Catalonia offers strong capabilities in genomics and structural biology, thanks to a solid research background, the leverage of a growing sector and a network of large infrastructures (Barcelona Supercomputing Centre, Alba Synchrotron and the National Genome Analysis Center). This supports top-ranked research centers performing activities ranging from basic research to industry collaborations. Catalonia excels in research and innovation in medical chemistry, "omic" sciences, bioengineering, crystallography and bioinformatics. Dynamic activity in the pharmaceutical and fine chemical industries, along with growing biotech and Bio-IT sectors, provide added value to research.

Catalonia has an outstanding network of science parks (16) and technological platforms (118), which provides an excellent meeting point for universities, businesses and society. Together, they can support high level applied science, fostering technology transfer, facilitating the creation of new companies and bridging the dialogue between science and society. Catalonia's capability to turn research into value facilitates the growth of local pharmaceutical, biotechnological and medical technology firms in one of the most complete translational medicine pipelines emerging in Europe for cancer, cardiovascular diseases and neurosciences. To strengthen these capabilities, Biocat promotes the establishment of collaborative networks within the private and the public sectors. Two examples are Oncocat (for cancer) and BioNanoMed Catalunya (for bionanomedicine). Both initiatives aim to raise awareness of Catalonia's potential in these arenas but, above all, to give companies and research centers a common meeting place. The two networks have more than 60 member companies and organizations.

This unique environment makes the BioRegion of Catalonia one of the main emerging clusters in Europe. Catalonia's science parks provide great opportunities for the incubation of biotechnology companies since they include scientific and technical infrastructure and business management consulting services.

14.1.3. Antecedents of Biocat

The initiator initiative arose from the research field in Catalonia in 2003: l'Aliança Biomèdica de Barcelona (Barcelona Biomedical Alliance). It was signed by the Institute of Biomedical Research August Pi i Sunyer (IDIBAPS), Barcelona Science Park (PCB) and the Barcelona Biomedical Research Park (PRBB) and promoted by DURSI, with the expectation of structuring and strengthening the quality and capacity of public biomedical research developed in Barcelona and its surrounding areas, while being projected as a mark of excellence in the field of international biomedical research. This alliance was based on a multicenter, highly coordinated model, and even then already, fully aligned with the goals of the new policy from the European Research and Innovation area to create excellence in the regional environments of science and technology, especially in biomedicine and biotechnology.

The mission of the Alliance, which also received support from the Ministry of Science and Technology of the State, was to be an entity to act as an interface between signatory biomedical research institutions, as well as participate with agents and other public and private parties in building a powerful biomedical and biotechnological center in the Barcelona area (Bio-cluster BCN).

The Research and Innovation Plan of Catalonia 2005–2008 was established in 2004. Biomedical research was identified, among others, as an area with great potential, and it was suggested that Catalonia could compete globally in this region. The Plan was the prime political instrument that would allow Catalonia to propel itself to the forefront of science in Europe, the novelty of which was its comprehensive approach and interdepartmental cooperation. The new plan revolved around three elements that involved the government's responsibility: research funding, participation in knowledge transfer and an evaluation of the performance.

Meanwhile, in the business environment, investment in R&D&I was lower than in the Europe's best-developed regions. Catalonia at that time showed some significant weaknesses (some from which it still suffers): a production structure with a majority of SMEs, lack of an innovation strategy and innovation management capabilities, low presence of Catalan companies in high-tech sectors, little entrepreneurial culture, lack of support structures for innovation and sectorial organizations and low participation from the business sector on policies to boost these structures. Regarding the qualified human capital (doctors, engineers and scientists), their inclusion in companies was still limited, which restricted the transfer of knowledge and technology.

In contrast to these weaknesses, the biotechnology sector in Catalonia showed a strong growth potential. New companies showed their ability to incorporate knowledge and continuously innovate. Catalan companies, at that moment, already had higher levels of participation than the Spanish average in the public calls for the promotion of research and development, both in the European Union framework programs and in the central state's programs. Catalonia had — and has — an important pharmaceutical sector, with six major domestic groups that accounted for 60% of the state pharmaceutical production.

New structures such as scientific parks also emerged, assisting the relations between the sector's parties, especially between universities and companies. This is the case of the PCB, established by the University of Barcelona in 1997, which was the first scientific park in

Spain. New economic agents also appeared at that stage, such as business angel associations and venture capital firms.

Given this rich and complex scenario, the industry perceived the need for structures to mediate between the different agents, particularly between the public research system and companies — in order to achieve the correct coordination of the research and innovation system and to promote the transfer of knowledge and technology. There were already some initiatives to boost technology transfer, such as the university TTOs, the IT network (a network of support centers for technological innovation), the network for technological springboards or the Knowledge Transfer Centre (CTC). However, an integrative sectorial action was still necessary and it needed the transversal support from the administration and all stakeholders. It also had to ensure a common strategic plan for all the Catalan biotechnology and biomedicine areas as a focus for the new economy based on knowledge and innovation. There was a need for Biocat.

14.1.4. Biocat's creation

In 2006, Biocat was launched. The role of Biocat was defined as the instrument which would bring together enterprises, public R&D&I institutions and public administration in order to promote the biotechnological sector in Catalonia not only in terms of research and innovation, but also in business creation and jobs. The Biocat universities, research centers, science parks, hospitals, governmental agencies and industrial partners joined in at the beginning.

Initially, Biocat was chaired by the Vice President of the Catalan government, Josep Bargalló. Then the foundation was presided over by the President of the Catalan government in the 2006–2010 legislatures in order to show the involvement of the government in this drive.

Biocat adopted the legal form of a foundation, and since its establishment has been governed by a board, a collegial body composed of representatives from the public administration and the private sector. In the first stage, 2006–2011, the Biocat executive bodies were the President of the Executive Board and the CEO.

Subsequently, a reform of the statutes of the foundation simplified the government structure, reducing it to a board (plenary delegate) and CEO. Since April 2007 Biocat's CEO is Dr. Montserrat Vendrell, who had taken part, from the PCB, in the conceptualization of the BioRegion of Catalonia. At the beginning of its establishment in 2006, the central government, the Catalan government and the Barcelona City Council assumed much of the Biocat funding. Later, this was supplemented by other sources such as direct participation in international competitive projects or sponsorship for some of the specific activities.

14.1.5. *The first strategic plan*

In late 2007 the first Biocat Strategic Plan 2007–2010 was approved on the basis that the first challenge was the configuration of the Catalan biotechnological and biomedical sector as a well-coordinated and competitive bioregion.

The first observation made was the multiplicity of parties involved in shaping a sectorial strategy for success. Furthermore, the assets of the Bioregion during the definition of the 2007–2010 Strategic Plan were unevenly distributed and the objective to build a strong and competitive Bioregion required the concerted action of several agents such as Biocat. Biocat proposed five strategic lines of action to organize their objectives and activities:

- Line 1: To make Catalonia an environment of international importance in biomedicine and biotechnology, attractive for business and industry researchers.
- Line 2: To consolidate Biocat as a one-stop shop in activities related to biotechnology and biomedicine in Catalonia.
- Line 3: To promote research in life sciences, as well as its valorization and marketing so that innovations reach the market.
- Line 4: To encourage the consolidation of the biotechnological and biomedical business sector.
- Line 5: To contribute to a better public perception of biotechnology, its benefits and its applications.

The percentage of achievement of the strategic objectives defined in 2007 was very high, reaching levels of 90–100% in most of the defined indicators, especially on lines 1, 2 and 4. The framework of strategic objectives remained stable during the period, although the set of actions was adapted — especially with regard to the implementation calendar — to a changing context, which since 2009 has been influenced by a deep economic crisis.

Early on, the OECD assessed the strategic role of Biocat as an adviser to economic and political decision-makers as important (OECD, 2010). This report also pointed out that the Catalan cluster "is among the five to ten top clusters in the world in regards to the dynamism of its network" (OECD, 2010).

14.1.6. *The second strategic plan*

A new Master Plan 2010–2013 that has guided the steps of Biocat these last four years was developed in 2010. The basic strategic axes are the result of a concentration on the key focal points for the promotion of competitiveness, internationalization and promotion and recruitment of talent, especially in the field of business management and entrepreneurship. This plan is further discussed below.

14.2. Identifying best practices

14.2.1. *Biocat's clear vision and mission*

Biocat's organizational structure has been put together in such a way that it can respond in the most effective manner to the mandate issued by the vision and the mission of Biocat.

14.2.1.1. *Vision*

Biocat faces the future as the organization leading a strong and well-structured biotechnological and biomedical cluster, knowing how to extract all the potential from the research capabilities of Catalonia, and promoting the its leadership within an innovative Europe.

14.2.1.2. Mission

Biocat's mission is to dynamize all the agents and initiatives involved in the biotechnology, biomedicine and medical technology sector in Catalonia, shaping an environment with a strong research system, active in technology transfer, and with an industrial tissue capable of becoming the economic driver of Catalonia and contributing to society's quality of life.

The actions of Biocat to realize its vision and mission are supported by its values: collaboration; flexibility and innovation; quality and efficacy; customer orientation; and commitment.

14.2.2. Governance

The board of trustees of the Bioregion of Catalonia Foundation (Biocat) is the top rung of the organizations and contains representatives from government (6), research institutions (4) and the private sector (13).

Biocat's team is organized into three functional departments: Communications and Institutional Relations; Business Development; and Finance and Services. The team is led by the Chief Executive Officer (CEO) who has an assistant (with the rank of Head of Department), responsible for the areas of Governance and Human Resources. The CEO and Heads of Departments form the Steering Committee of Biocat.

This organization responds to a conceptual schema that places the leadership of a strategic role in the organization in each area (Figure 14-1).

14.2.3. Core business

The Department of Business Development is located in the "core business" of the organization, designing and managing programs that respond to the needs of the various entities in the cluster, throughout the different stages of the value chain, from research to market access and internationalization.

14.2.3.1. Visibility/promotion

The Department of Communications and Institutional Relations plans the communication strategy and manages all the tools

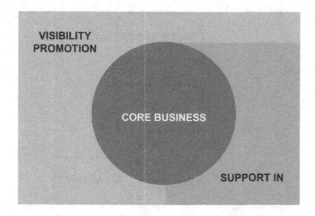

Figure 14-1: Strategic roles of the Biocat organization.

for visibility and promotion in addition to those for managing knowledge and the relational capacity of the organization. The CEO plays a key role in the institutional relations and communication strategy, catalyzing the visibility and public representation of the organization.

14.2.3.2. *Support in*

The Department of Finance and Services is responsible for the basic tools and strategies for the management of the organization (administration and accounting systems) together with CEO's assistant, who is responsible for human resources. The CEO's assistant is also responsible for relations with legal counsel and entities linked to the governance of Biocat (public and private institutions with representation on the Board). The Biocat governance structure is shown in the following Figure 14-2.

14.2.4. *Main activities*

All of Biocat's programs and activities are structured around four strategic axes: (1) cluster consolidation; (2) business competitiveness and talent; (3) internationalization; and (4) social perception of biotechnology. Next to this, there are two crosscutting

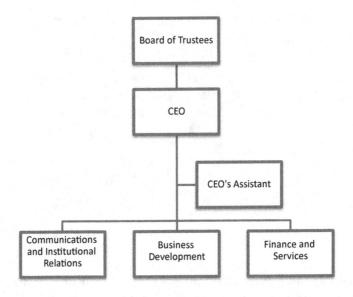

Figure 14-2: Biocat governance structure.

strategic projects that impact each axis: KIC Health and Food and B·Debate.

The activities carried out by Biocat since its creation and the results obtained can be consulted in the archives of the organization (http://www.biocat.cat/es/publicaciones/memoria), available since 2006 in Catalan, Spanish and English.

14.2.4.1. Cluster consolidation

From the beginning, one of Biocat's priorities has been to identify the members and stakeholders in the Catalan bio-cluster and foster networking and synergies among them. The members of the organization participate in an average of 50/60 external public events per year. The region's consolidation as a network of knowledge and collaboration is essential to achieving goals related to both scientific and business growth and improvement. Some of the tools and programs that are included in this axis are:

> Biocat Directory: more than 1,200 profiles provide basic information on all the stakeholders in the sector, classified by subsectors, areas of

activity and therapeutic specialization. This online tool is available at all times (in English) on the Biocat website.

Biocat Report: a biannual study focusing on the state of biotechnology, biomedicine and medical technology in Catalonia (http://www.biocat.cat/en/publications/report).

Creating and driving specialized networks: the BioNanoMed Catalonia (www.bionanomedcat.org) and Oncocat (www.oncocat.org) networks bundle research centers, hospitals and companies that carry out similar research.

Publications, databases, analyses, opinions and news find their space on Biocat's website (www.biocat.cat): the website has an average of 20,000 monthly visits and 17,000 unique visitors, and has recently been recognized as the second best web of the European bioclusters (Bio Web Score created by Biotech Finances electronic magazine; http://www.biotech-finances.com/en).

Sectorial events and meetings like the Biocat Forum or Biocat workshops where the future challenges facing stakeholders in the BioRegion are discussed. Fòrum Biocat gathered between 500 and 700 participants from all stakeholders. Nearly 2,500 professionals, researchers, entrepreneurs and public administrators have participated in more than 30 sectorial conferences.

14.2.4.2. Business competitiveness and talent

Fostering entrepreneurial spirit and driving the creation of new companies through increased technology transfer is key to the BioRegion's development. Biocat team members offer personalized advice to entrepreneurs and companies (about 150 meetings/year). This advisory work complements the contacts offered with a number of international advisors (http://www.biocat.cat/en/profile-advisors).

The consolidation and growth of existing companies is equally important to Biocat. Biocat plays a key role in interactions between investors and biotech and techmed companies, helping identify opportunities and participating in the evaluation of innovative projects. It also facilitates the creation of business consortia for large-scale

R&D projects. To face these challenges, Biocat has the following programs and tools:

> International advisory services: Upon analyzing Catalan biotechnology and medical technology companies, Biocat has detected four priority areas for improvement: strengthening strategic management bodies (boards of administration); business development (market analysis and positioning); R&D strategy; and intellectual property portfolio management (patents). In order to address these weaknesses, Biocat offers the sector the expertise of its members and, above all, its network of contacts with national and international experts.
>
> BioEmprenedorXXI: through a telematics platform and various sessions, this program provides support for participating entrepreneurs with training, tutoring and mentoring activities, as well as facilitating their participation in financing forums (see also Case Study 12 on BioEmprenedorXXI). In its first five years of operation (2008–2013), 95 projects participated in the program, generating 51 new companies (biotechnological and medical technology).
>
> Technological offers and requests platform: Biocat is a member of the Enterprise Europe Network (EEN), which is present in more than 50 countries and offers a free tool to help European companies locate international technology partners.
>
> Talent for competitiveness program (2009–2010): this program, which was carried out in conjunction with the Department of Economics and Finance of the Generalitat de Catalunya in 2009 and 2010, was intended to strengthen the competitiveness and internationalization of Catalan companies and facilitating, by direct grants of up to 20,000 euros per company, the timely hiring of strategic experts in areas such as market exploration and business development, legal strategy and the strategy of R&D. In two terms of the program a total of 34 companies were benefited, receiving payments totaling 450,000 euros.

Additionally, taking into account that developing and attracting talent is key to driving competitiveness in the sector, we offer the sector training actions such as the Biocapsules program (20 training programs, 500 participants) or the Summer School on Medicines.

14.2.4.3. Internationalization

From basic research through production and commercialization of new drugs or innovative medical devices, the entire biotech sector moves in the international arena and the capacity of Catalan companies to act and compete in this environment is key to their development and future viability. One of Biocat's priorities is therefore to foster the presence and activity of Catalan companies in the sector in strategic markets with a focus on emerging economies, while promoting cross-border collaborations and projects to improve international scientific cooperation, technology transfer and access to the markets. As a basis for international action to improve our companies, we will improve communication tools, market analysis and coordination with other key stakeholders (ACC1Ó, Cambra, etc.).

To this end, the BioRegion of Catalonia participates in many international networks. Additionally, Biocat also participates in European transnational cooperation projects to foster market access and technology transfer (ETTBio, Transbio, etc.). Through this network of international relations, Biocat maintains close knowledge links to countries like Germany, Belgium, Brazil, Canada, China, the United States, France, the Netherlands, Hungary, India, Poland, the United Kingdom, Sweden and Switzerland.

Biocat also promotes and coordinates the participation of Catalan companies in international conventions, partnering events and fairs such as BIO and BIO-Europe. In total, 78 Catalan biotechnology companies participated in various editions of the BIO fair, constituting between 50% and 60% of the Spanish. Between 2006 and 2013, Biocat has organized a total of 45 international missions. A second line of work focuses on attracting and organizing international events to be held in Catalonia, and more specifically in Barcelona.

14.2.4.4. Social perception of biotechnology

The lack of information and understanding of breakthroughs in the life sciences and their applications sometimes causes confusion and can undermine appreciation of the benefits they contribute to our quality of life. Therefore, one of Biocat's strategic priorities is to raise awareness of biotechnology in Catalan society, from scientific research to applications for creating and developing new products (drugs and therapies, medical technology, food products, biofuels, industrial bioprocesses, materials to fight contamination, etc.).

14.2.4.5. KIC-IET

The EIT is a body of the European Union whose mission is to increase European sustainable growth and competitiveness by reinforcing the innovation capacity of the EU. Biocat, with the support of the Catalan Government and the State Government, decided to promote a dual candidacy in direct relation to health. The first, a proposal to focus on innovation to an active lifestyle and healthy aging and the other focused on feeding the future.

Biocat leads all these initiatives related to the EIT carried out in Catalonia along with its different partners, in order to achieve the objective of having a headquarters in case of being selected as a future KIC.

14.2.4.6. B·DEBATE

B·Debate — International Center for Scientific Debate Barcelona is a joint initiative of Biocat and the "la Caixa" Foundation, which began its activity in 2009. The main objective of B·Debate is to promote top-notch scientific debates and meetings for the open exchange of knowledge and collaboration among experts of renowned international prestige in order to tackle complex social challenges in the life sciences arena.

14.3. Implementing best practices in your region

This case study highlights several guidelines for stakeholders or policy makers in bio-regions or clusters that wish to build a cluster organization. It is clear that founding and growing an organization such as Biocat requires a number of key ingredients. It is clear that there were already extensive activities in the field of biotechnology

R&D&I, in the academic as well as in the industrial sector. Driven by the government, a platform was created to bring all stakeholders together and representatives from all sectors are present in the board of trustees. Start-up funding was provided by the Catalan government but this funding was quickly complemented by funding from other, external sources. Biocat is organized in three main departments, each responsible for a key functional domain of a cluster organization: communication, business development and support. This clear organization makes that Biocat can work around four strategic axes. All Biocat activities fit within the four strategic domains and are characterized by a number of transversal principles/success factors:

- Collaborate, collaborate, collaborate: Biocat strongly believes in collaboration, not only between its member organizations but between all organizations and individuals trying to advance the biotechnology scene.
- Go global: in the biotechnology industry or research environment, staying local is not an option. Biocat involves international organizations and advisors in all its activities and network. Tapping into this knowledge is crucial for the future success of a bio-cluster.
- Educate: tapping into knowledge is one thing, keeping it in the region is another. Biocat organizes a lot of trainings and educational events for its researchers, entrepreneurs, managers and policy makers. Only a high concentration of world-class knowledge can ascertain long-term competitive advantage for this reason.
- Be proud: achievements should be carefully communicated and used to enthuse all stakeholders of the cluster.
- The success of the cluster is the success of Biocat.

14.4. Future opportunities

Biocat and other stakeholders are currently analyzing future opportunities under a new strategic approach.

14.5. Best practices

- A clear vision and mission, supported by all life sciences stakeholders in the region.
- Strong government support from inception.
- Management structure centered around three main functional domains.
- Clear definition of four strategic axes and all actions and programs fit in the strategic axes.
- Strong focus on international networks, communication and education.
- Convincingly branding the Catalonian bio-cluster.

14.6. References

Asebio Report (2013). INFORME 2012 ASEBIO: Situación y tendencias del sector de la biotecnología en España.

Biocat Report (2013). 2013 Biotech Report: Catalan Life Sciences — Status and Analysis: Commited to Value Driven Growth. Barcelona.

OECD (2010). *Reviews of Regional Innovation: Catalonia, Spain.* OECD, Paris, France.

15

CASE STUDY 15: The Effects of a Cluster on a Spin-Off — The Foundation of Ablynx

Tom Guldemont, Thomas Crispeels and Ilse Scheerlinck
(Vrije Universiteit Brussel, Vesalius College, Belgium)

15.1. Setting the scene

In order to illustrate the importance of networks and clustering in the domain of life sciences technology transfer, we present a comprehensive case study on a particular success story: Ablynx. Currently, Ablynx is a stock quoted spin-off company of the VrijeUniversiteitBrussel (VUB) and the Flanders Institute for Biotechnology (VIB). Its main activities are the development of novel therapeutics based on a proprietary technology platform — the nanobodies — and the company employs more than 250 professionals.

This case study draws on a number of interviews of and guest lectures by the people closely involved in the start-up of Ablynx and the technology transfer activities preceding the foundation of the company such as the inventor, tech transfer officer and investor.

The origins of the Ablynx story can be traced back to the biotechnology laboratories of Professor R. Hamers at the VUB at the end of the eighties. This laboratory was a melting pot of different scientific disciplines and scientists. During the seventies and eighties, Belgian biotechnology was well developed under impulse of world-class researchers and pioneers such as Walter Fiers, Jeff Schell, Marc Van Montagu, Desire Collen and Raymond Hamers. The Hamers lab consisted of several research groups that were active in a wide variety of fields: a.o. molecular immunology and parasitology, cellular and DNA-protein interactions. In the latter group, a research project/group was trying to generate monoclonal antibodies targeting DNA-molecules. At a certain point in time, it was decided that a joint research project, involving the three main groups in the lab, should focus on the role of carbohydrates in cell interactions. One of the activities in this joint project was to try and generate monoclonal antibodies targeting specific carbohydrates. This effort was headed by Professor S. Muyldermans.

All interviewees describe the head of the department, Professor Dr. R. Hamers, a scientist with a very broad knowledge and field of interest. He was a biologist, but could discuss with any scientist on any chosen topic. He was very active in attracting and inviting students and scientists from around the world to the laboratory. His efforts to bring scientists together and engage in many research projects led to a shortage of financial means at that time among the drivers to engage in technology transfer and spin-off creation.

Meanwhile, a lot of things were moving in the region concerning biotechnology. Flanders, the Northern part of Belgium, was a center of academic and commercial biotechnology excellence. This is illustrated by the work of researchers like Marc Van Montagu and Jeff Schell who invented the first technology to introduce recombinant genes in plants. Shortly after this breakthrough, one of the world's first agricultural biotechnology companies, Plant Genetic Systems (now Bayer CropScience), was founded in Ghent in 1982. Simultaneously, Walter Fiers, Desiré Collen and many other Belgian researchers became renowned for their advancements in the field of biotechnology and together with the foundation of companies like Innogenetics and Thromb-X, these breakthroughs illustrate the biotechnology excellence present in the region at that time. The presence of these early successes by pioneering companies will prove to be very valuable for the rest of the story since they form the seedbed of Ablynx. For instance, the author of the initial business plan, Gaston

Matthyssens, and the first CEO of Ablynx, Mark Vaeck, both worked in the laboratory of Raymond Hamers and at Plant Genetic Systems during the eighties.

15.2. Identifying best practices

15.2.1. *Trigger*

The discovery of the nanobodies can be traced back to a practicum in immunology taught by Professor Hamers at the VUB. During this practicum, students in bio-engineering were challenged to isolate antibodies from human blood. This blood was to be obtained from one of the teaching assistants. But as these were the late eighties the students were reluctant to work with human blood that was not tested for HIV contamination. The teaching staff then told the students to work with mouse blood instead, like any other student in the world. Once again, this idea was challenged by the students as being very unoriginal and quite pointless since the experiment had been done for years by scientist all over the world. At that point, Professor Hamers dug up samples of camel blood he had received from a laboratory in Mali. These samples had been sent to Brussels in order to be tested for the presence of trypanosomes, a parasite. It was proposed that these samples would be used in the practicum. The students started working on isolating antibodies from camel blood, an experiment that no one had done before. The students returned with unexpected results. Professor Muyldermans recalls the day the students presented him the results: "There were three possibilities. The first was that the students had done something wrong. A second possibility was that the blood samples we received from Mali had degraded during transport or storage. A third possibility was that there was really something strange going on in camels' blood." At this point, two possible ways of proceeding were considered. Either the results were discarded, or further research would have to be done to resolve the issue. The scientists, Professor Hamers first, opted for the latter solution. A team of researchers drove to the Antwerp Zoo in order to collect a new, fresh blood sample from a camel. The results were the same as in the previous experiment and a very interesting hypothesis was built: it seemed that camelids possessed a different class of antibodies, next to the classic antibodies. Antibodies are proteins that have a specific function in the immune system of animals: they bind to antigens (foreign bodies) in the body and induce an immune response. A classic antibody, as can be found in humans, has a functional domain (i.e. the site that binds to the antigen) that consists of two polypeptide chains. Some antibodies in camelids however have a functional domain consisting only of a single polypeptide chain. This had never been observed before.

The value of these nanobodies, though not yet demonstrated, was almost immediately spotted by the researchers. While the whole scientific community was struggling to produce classic monoclonal antibodies[5] and their fragments — which were unstable, difficult to clone and hard to produce (mammalian cell line was needed needed) — the researchers at the lab of Hamers were working with the so-called nanobodies which were easy to clone, very stable and easy to produce (can be produced in lower eukaryotes).

15.2.2. On patents and publications

The abovementioned constant shortage of financial means in the lab meant that researchers were constantly looking for opportunities to fund their research. This, among others things, led to the filing of a patent application on the nanobodies. Although there were already some commercial "success stories" present in the Flanders biotechnology industry, the idea of an academic researcher patenting his work and pursuing a valorization route was novel in the early nineties in Europe and certainly at the VUB. Due to the vision and persistence of the people at the laboratory, a patent application on the

---

[5] See for instance the time between invention of hybridomas by Kohler and Milstein (1975), Start-up of Hybritech (1978) and first commercially available antibody therapy Herceptin (Genentech, 1996). Large contrast with first recombinant human protein therapies on the market (less than 10 years after development of recombinant DNA technology).

nanobodies was filed by the three inventors (interestingly, not by the university).

It should be noted that during these days, academic researchers were under less pressure to publish their results as they are nowadays. This allowed the academic staff to first investigate and elaborate the discovery and possible technologies emerging from it. In this way, the lab could obtain a head start on competing research groups that would pick up the technology quickly when the results were published. Also, the extended stay of the technology within the walls increased the body of knowledge on the technology and thus increased the chances of achieving a top publication. Luckily, this strategy meant that the discovery was not disclosed to the public before more knowledge was generated. This allowed the patenting of the technology.

When Professor Hamers approached the VUB administration with the request of supporting his patent application, the central university administration had no experience with these matters. In Europe, technology transfer was not considered to be a core function of a university. This was different in the US where the Bay-Dohle act (1980) triggered the rise of technology transfer offices at universities.[6] The links between industrial and academic world were scarce and weak. After his visit to the university administration, Professor Hamers decided to file for patent on his own and contacted an experienced patent attorney in Paris. Together with Casterman, Muyldermans, Frenken and Verrips, the co-inventors, he filed for a patent on their discovery in the summer of 1992, without any backing of the university. Being the first to observe antibodies devoid of their light chains, the inventors were able to file a product claim (instead of the more common method claim). This means that the patent, if granted, would cover any product that contained or resembled a nanobody. This very broad scope is nowadays rare in biotechnology but gave the patent applicants a broad patent protection, which is seldom seen in therapeutic biotechnology.

---

[6] A Belgian equivalent for the Bay-Dohle act was only implemented in 1998.

15.2.3. The top publication

Although the first public disclosure of the results happened in December 1992 at a scientific congress when the scientists presented a poster, the big breakthrough publication was published nearly a year later. In June 1993 the article "Naturally Occurring Antibodies Devoid of Light Chains" was published in *Nature* (Hamers-Casterman *et al.*, 1993).

Meanwhile, the cost for maintaining the patent application was increasing and the scientists were paying for all of this with their own money. So, once again, they turn to the university administration and asked for their help. The university still had no internal ruling or legal framework, nor did it have a technology transfer cell. But the publication of the results in *Nature*, the strength of the intellectual property protection and the fact that competitive actions were undertaken by a large company made it clear to the university that the technology held considerable valorization potential. In 1995, they reached an agreement on transfer of the patent to the VUB and distribution of possible income from the patents. These negotiations with Professor Hamers and his team were one of the first steps of the young technology transfer cell of the VUB. In parallel with these developments, a new important player had emerged in the regional biotechnology community: VIB. This research institute would prove to be extremely instrumental in building a company from the nanobody technology. In the following section, we briefly discuss the history and role of this institute.

15.2.4. Government intervention

A couple of years prior to the discovery of the nanobody technology, the Flemish Government had launched the VLAB-program (Vlaams Actieplan Biotechnologie, Flemish Action Plan Biotechnology), which was specifically designed to detect and support promising biotechnology projects. Remarkably this is the focus of the agency towards valorization. Rudy Dekeyser and his colleague Jo Bury were both scientific consultants at the Agency for Innovation through Science and Technology (IWT) and responsible for the VLAB, later

VLAB-ETC, program. In 1993, they were both invited to the office of Luc Van den Brande, then Minister-President of the Flemish government. Wanting to put the Flemish biotechnology industry on the world map, Van den Brande asked them to provide recommendations as well as a plan to accomplish it.

> He said: "I want to invest 1 billion Belgian francs [25 million euros] annually to put the Flemish biotechnology on the world map. Excellence in research has to be the focus, but the translation to social and economic value is at least as important. Here is a white sheet of paper, to you the task is to forge a plan." (VIB, 2012)

Dekeyser and Bury immediately had a new "institution" in mind; they didn't merely want to create a new, biotechnology-specific, "granting body" because this approach held the risk of spreading the government funding too thin. Together with government employee Dirk Callaerts, Bart De Moor and Christine Claus, Director General of IWT, Jo Bury and Rudy Dekeyser formed a team that gradually elaborated on the plan. In February 1994, less than ten weeks after their first meeting, Dekeyser and Bury returned to the Minister-President's office. The first plan was unfolded, and Van den Brande saw that it was good. He sent Dekeyser and Bury — discretely — to the field to check the reaction of stakeholders to the plan (VIB, 2013). Some were very enthusiastic, others responded much more cautious. In April 1994, the plan was officially presented and discussed in the parliament and the Flemish government. The period between April 1994 and April 1995 was therefore quite bumpy and littered with many discussions. One of the main concerns of the academic community was that only two of the four Flemish universities would be present in this new biotechnology structure/institute.

On April 5, 1995, during its last meeting, the Flemish government approved the establishment of the VIB. The new initiative was a research institute with four core departments, more specifically the labs from Herman VandenBerghe and Desire Collen (both KULeuven), and Walter Fiers and Marc Van Montagu (both Ghent University). These core departments were supplemented with five associated departments: the labs of Nicolas Glansdorff (Vrije Universiteit Brussel),

Raymond Hamers (Vrije Universiteit Brussel), Danny Huylebroeck (KULeuven), Christine Van Broeckhoven (University of Antwerp) and Joël Vandekerckhove (Ghent University). In fact, the nanobody know-how seemed to be crucial in order to incorporate a VUB research department in the VIB. VIB became an institute without walls where the research groups remained on their university campuses. To enable this, framework agreements were signed between all universities involved and VIB. VIB also set up an administrative headquarters in Ghent. To achieve the objectives, the Flemish government provided an annual grant of approximately 23 million euros to the institute (VIB, 2013).

At inception on January 1, 1996, and after flipping a coin, Jo Bury became Managing Director of VIB, while Rudy Dekeyser became Deputy Director and Director of Technology Transfer.

15.2.5. The nanobody story continues

When the VIB was finally established, the IP rights regarding the nanobodies were transferred from VUB to the VIB. More particularly, VUB remained the owner of the patents, and granted an exclusive license on the patents to VIB.

The nanobody technology was further developed in the labs of Professors Lode Wyns, Jan Steyaert and Serge Muyldermans (VUB), but now under the flag of the VIB. In particular, the development of the PiCup proteomics platform by professor Steyaert and the following proof-of-concept studies triggered the interest of venture capitals.

In 1996, the nanobodies team was able to attract an IWT-SBO grant. In fact, it was the first SBO-project in Flanders (SBO stands for strategisch basisonderzoek, which translates to strategic fundamental research). This subsidy was actually already granted to the team in 1994, but was revoked in the beginning of 1996. In the meantime however, the VIB was established, with the nanobodies group under its wings. Thanks to efforts of VIB at the address of IWT, the IWT subsidy was re-activated a few weeks later and made an essential contribution to the continuation of the story. A subsidy for an industrial development program, the IWT grant was equivalent to seed funding. The considerable budget of 1.9 million euros enabled the research group to attract young scientists that later would become the Ablynx core research team. These young researchers had the triple challenge to develop novel nanobody-based technologies, to strengthen the patent portfolio and to develop proof-of-concept examples to show the strength and competitiveness of the technology.

15.2.6. Establishment of the company

A couple of years and three versions of the business plan later, Ablynx was founded on 15 July 2001. The company was set up with a minimum of capital (less than 100,000 euros) provided by GIMV (Venture Capitalist), VIB, Biotech Fonds Vlaanderen and Mark Vaeck (CEO).

Mark Vaeck, an experienced biotech entrepreneur and one of the initial investors, joined as CEO and was recruited through the network of GIMV. Gaston Matthyssens, at that time business developer at VIB, wrote the initial three versions of the business plan and Vaeck reformatted the last version of the business plan in collaboration with VIB and the new investors.

In November 2001 Ablynx reached an agreement with VIB on the transfer of the nanobody technology and IP. That is, VIB granted an exclusive sublicense on the nanobody to Ablynx for healthcare applications. In addition, the first financing round of 2 million euros by GIMV was closed. In August 2002 follow-up financing of 3 million euros was provided by Sofinnova (FR) and Gilde (NL), which resulted in a total round A financing of 5 million euros.

The start-up team was hired during the Christmas period of 2001. Five scientists from the Hamers' research group (VUB/VIB) joined with the first objective to transfer the technology and know-how and to prove the industrial potential of the technology.

In November 2007 Ablynx successfully completed its initial public offering (IPO) and is now listed on Euronext Brussels under the symbol ABLX. Up until now, Ablynx raised a total of 200 million euros from both private and public investors. It has more than 250 employees and the exclusive rights to more than 500 patent applications and granted patents. Approximately 25 programs are in their R&D pipeline. Two products achieved clinical proof of concept in rheumatoid arthritis, while five nanobody products are in the clinic, three in Phase II and two in Phase I. Additionally, the company has partnerships with Abbvie, Boehringer Ingelheim, Merck Serono, Novartis and Merck & Co.

15.2.7. *A story of people*

Building up value that led to the foundation of Ablynx required some basic ingredients:

- A patented, differentiating technology platform: Nanobody® technology, protected by the Hamers patents.
- Technology transfer mechanisms: the tech transfer team of the VIB was in the driver seat, with major contributions made by the VUB team, the scientists of the lab and the government (IWT).
- The presence of experienced investors (in the region): GIMV, Sofinnova and Gilde.
- A serial entrepreneur for CEO: Mark Vaeck.
- A highly motivated scientific team: initial science team from the Hamers group.

One of the striking observations in this case is the fact that a large number of the Ablynx stakeholders, at least the ones we interviewed, had a relationship prior to the emergence of the nanobody/Ablynx project. Maarten Sileghem, the project champion at the IWT is a VUB-alumnus and was trained in the lab of Professor Hamers by Mark Vaeck (the first CEO of Ablynx) and Serge Muyldermans (nanobody co-inventor). Gaston Matthyssens, the business developer of the VIB at the time of project incubation, who authored the first versions of the business plan, had also been a researcher at the lab of Professor Hamers. Rudy Dekeyser had been active in the local biotechnology research and business community as manager of the

VLAB and returns to the story when the intellectual property rights are transferred from the VUB to the VIB.[7]

15.3. Implementing best practices in your region

15.3.1. Implications for TTO

Support visionary scientists that want to commercialize their invention. This recommendation may seem obvious and straightforward, however, putting this recommendation into practice is a lot harder. In the beginning of the story, the research team did not receive tech transfer support of the parent institute because it was considered inappropriate and there were no resources to do it. Luckily, awareness of technology transfer and the means flowing to TTOs allows for most institutions to provide some form of support. It is clear however, that a lot of time was lost between discovery and foundation of the company due to inexperience and a lack of suitable structures and processes. The main message of this story is to support and believe in excellent science and that excellent valorization opportunities will follow automatically.

15.3.2. Implications for universities/research organizations

Foster multidisciplinary research. As mentioned previously, Professor Hamers' lab was a melting pot of different branches of science, something that has become hard to maintain under pressure of the extreme specialization going on in the scientific world. Recognizing and rewarding true interdisciplinary research is key as interesting things always happen in such research.

15.3.3. Implications for policy makers

Be visionary, think long term. Long-term commitment in supporting biotechnology and its valorization. Free up budget to set up

---

[7] This list is not exhaustive and we acknowledge the contributions of other stakeholders, investors and scientists. It would be impossible to include/mention them all.

initiatives like VLAB, IWT and VIB, earmarked with technology transfer objectives. Provide funding, not only for research, but also for IP and valorization. Consider not spreading the funding for research over all research groups, and supporting top research groups in a structural way.

Attract and invest in talented people and support a cluster/network organization; observe that scientists go abroad/to the industry, but return. Stay in touch.

15.3.4. *Implications for scientists*

The cliché quote from Louis Pasteur, "Chance favors the prepared mind", definitely applies to this case study and scientists involved were actively looking beyond the boundaries of their project and even scientific discipline.

15.4. Future opportunities

The main limitation is the fact that you need a combination of ingredients to have a success story. It is often an AND-AND story: top research AND top people AND governmental support AND good tech transfer policy. If one of them is missing, chances of success in terms of valorization will decrease. This also means that it takes a lot of time if you have to start really from the beginning. Especially the top research part is something you cannot influence or accelerate in the short term (e.g. this already started in the seventies in Flanders).

It's rather hard to improve a story that occurred in the past. What can still be done is trying to maintain and further improve networks and relationships, in order to enable organic growth of the sector. Therefore, it is important to have a good cluster organization that can take care of that.

15.5. Best practices

- Top visionary people, not afraid to swim against the stream.
- Top research: starting in the seventies in the already existing university labs, but continued and further stimulated by the VIB.

- Forward looking government: initiatives like the VLAB, IWT and VIB have had a crucial role in the start-up of Ablynx and many other spin-offs. The Ablynx story also demonstrates that new elections can have a catalyzing effect on these initiatives.
- The combination of the previous three strengths in the beginning of the nineties made the whole community want to go in the same direction. All necessary ingredients were available, for a large part spinning out of the early commercial successes of the eighties, which really proved to be a seedbed for later developments: top research, top people and government support. In other words, the momentum was there.
- In the beginning of the nineties there was less publication pressure, which has proven that filing for a patent and a publication can go hand in hand.
- The Ablynx story triggered the first technology transfer activities and experience of the VUB. It proved to be the germ for what later would become the Technology Transfer Interface of VUB.
- All important people in the early history of the nanobodies and Ablynx were in some way connected to each other. This makes it easier to pick up the phone when you have a question instead of arranging formal time-consuming meetings, allowing swift decisions. It highlights the importance of a strong network.

15.6. References

Belgian Foreign Trade Agency (2011). Belgian Biotechnology. Retrieved from: http://flandersbio.be/files/BIOTECH_WEB.pdf. Accessed on 15.06.2013.

Bulens, F. (2008). The Ablynx Case [Presentation]. Vrije Universiteit Brussel, Brussels, 21.05.2008.

Flandersbio (s.d.). 10 Reasons to set up a Life Sciences R&D entity in Flanders, Belgium. Retrieved from: http://flandersbio.be/files/Factsheet_FINAAL.pdf

Flandersbio (2011). Biotech Guide Flanders 2011. Retrieved from: http://flandersbio.be/downloads/. Accessed on 01.07.2013.

Flandersbio (2012). FlandersBio Activity Report 2011. Retrieved from: http://flandersbio.be/files/FlandersBio_activiteiten2011_lores.pdf. Accessed on: 02.07.2013.

Hamers-Casterman, C., Atarhouch, T., Muyldermans, S., Robinson, G., Hamers, C., Songa, E.B., Bendahman, N. & Hammers, R. (1993). Naturally occurring antibodies devoid of light chains. *Nature*, 363(6428), 446–448.

Matthyssen, G. (2008). *The initial Ablynx business plan* [Presentation]. Vrije Universiteit Brussel, Brussels, 23.04.2008.

Moses, E. (2013). Ablynx [Powerpoint Presentation]. Knowledge for Growth 2013, Ghent, Belgium, 30.05.2013.

Muyldermans, S. & Haesen, S. (2008). *The nanobodies at the Vrije Universiteit Brussel* [Presentation]. Vrije Universiteit Brussel, Brussels, 22.04.2008.

Vaeck, M. (2008). The *Ablynx Case: Starting and Building the Company* [Presentation]. Vrije Universiteit Brussel, Brussels, 7.05.2008.

VIB (2013). VIB renews General Management — Interview with Rudy Dekeyser. Retrieved from: http://www.vib.be/en/news/Pages/VIB-renews-General-Management---Interview-with-Rudy-Dekeyser.aspx. Accessed on: 06.06.2013.

VIB (2013). History of VIB. Retrieved from: http://www.vib.be/en/about-vib/organization/Pages/History.aspx. Accessed on: 04.06.2013.

16

CASE STUDY 16: Brokerage Event: Matching International R&D Projects

Regional Development Agency Ostrava
(RDAO, Czech Republic)

16.1. Setting the scene

The Brokerage Event (BE) is an event where researchers from research institutions and companies interested in sharing new project ideas and finding collaboration partners meet and discuss. The event offers the combination of an international fair, a conference and bilateral meetings. Participants can meet a lot of potential project or business partners in a short time, and meet a number of key players from familiar networks and projects. The participants can also meet international researchers and encounter the most recent technologies and applications.

According to the Regional Development Agency Ostrava (RDAO) (2012a), the Moravian-Silesian Region has a long-standing tradition in heavy industry — mechanical engineering, metallurgy and mining — which still has a substantial influence on the region's character today. As a result, the majority of the research and development capacities in the region are associated with big companies in these industries. There are other promising domains such as IT, electronics, biotechnology and environmental technologies of which R&D activities are primarily concentrated at the VŠB (Technical University of Ostrava, University of Ostrava and University Hospital Ostrava) as well as at several companies involved in the development of new products and technologies. With regard to public research and development, the VŠB-TU Ostrava plays a unique role: it is currently implementing several projects of new research centers, supported by the Operational Programme Research and Development for Innovation (a total of 160 million euros).

Besides the aforementioned universities, the Moravian-Silesian Region is the seat of other public universities and colleges (the Silesian University in Opava, College of Social and Administrative Affairs in Havířov, and Business School Ostrava), attended together by approximately 40,000 students every year. Most of the research sites and colleges were established in response to the economic development which followed the changes and restructuring waves in the latter half of the 20th century. Interestingly, the most important industrial sectors in the Moravian-Silesian Region are already organized in clusters, which gave the Region a new profile and made it much easier for investors to access individual subcontractors. In this respect, the Moravian-Silesian Region is the leading Region in the Czech Republic and harbors ten functional clusters in the Moravian-Silesian Region (RDAO, 2012b).

The Moravian-Silesian Region also responds to the current trends in the world economy and related approaches of regional economic development. In order to achieve successful development of regional economy, regional policy makers intensify investment in public and private resources for R&D activities with an emphasis on their effective transformation into innovative products. The focus of these efforts lies on industries traditionally characterized by high growth potential (e.g. biotechnology, environmental technologies, specialized services, design etc.) (RDAO, 2012c).

The field of biotechnology is considered one of the region's most promising industries, as indicated in the regional strategic documents. There are several measures that can be installed to support these new industries. One of the most important domains of support is the internationalization of companies and knowledge institutions, since in today's globalized world it is not possible to carry out research, development and transfer of technology in isolation. The Regional Development Agency Ostrava realizes this issue and within the creation of the main plan for the development of innovative activities in the Region (Regional Innovation Strategy of the Moravian-Silesian Region 2010–2020) (RDAO, 2012a), the agency carried out a detailed survey with companies and research institutions (including universities). The results of this survey comprehensively describe the current status of activities in research, development and innovation in the Region, including a comparison with other regions within the Czech Republic. The results of the survey reveal that the main obstacles for collaboration among research institutions and firms are characterized by a lack of own funds, a lack of own researchers and distrust of foreign partners. However, there are already first seeds of cooperation among entities in the field of biotechnology. Future collaboration is based on the contacts and acquaintances that arise most often from participation in conferences, seminars or workshops. It happens only rarely that cooperation arises in response to published results of research or on the basis of information gained from the public databases.

For this reason, the Regional Development Agency Ostrava, as the responsible authority for the implementation of the Regional Innovation Strategy, proposed measures that focus on creating and enhancing the opportunities to establish international contacts and on transferring know-how across organizations. This is achieved by organizing an international event focused on the exchange of experiences and knowledge: The Brokerage Event.

16.2. Identifying good practices

The Brokerage Event is a specialized conference focused on the presentation of up-to-date results of R&D within a specified sector and that hold potential for commercialization. It aims to bring together those organizations that have innovative solutions, can help develop a new product or improve manufacturing processes. The event is based on the concept of Open Innovation, which recognizes that the results of scientific practice can be found outside the organization. The target groups of these events are companies, universities, research institutes and public institutions supporting research and development from the Moravian-Silesian Region or from other regions of the Czech Republic or from abroad. The objective of the Brokerage Event is to provide space for international exchange of experience and knowledge on concrete research issues. Based on presentations and bilateral or multilateral negotiations, the event generates new international research and development projects that have the potential to transfer their results into practice.

The organizer of the Brokerage Event is the Regional Development Agency Ostrava (RDAO) in collaboration with selected partners from the region. Implementation of the Brokerage Event is also embodied in the Regional Innovation Strategy of Moravian-Silesian Region, where RDAO is also in charge of many activities. The procedure to set up a Brokerage Event is as follows:

1. RDAO proposes a topic, based on its expertise and on the needs of the region.
2. RDAO finds strategic partners in the region among companies and research institutions that are specialized in the field of interest and that have links with foreign partners.
3. RDAO, in collaboration with the partners, specifies the program and agenda of the event.
4. Strategic partners also help to attract speakers and international experts to the event.

The event is funded by the organizer (or more precisely by Regional Authority), strategic partners and through sponsorships

16.2.1. *Evolution of the Brokerage Event*

16.2.1.1. *The Brokerage Event 2011*

Ostrava's first ever Brokerage Event was an international conference focusing on R&D in biomedicine and on innovation in environmental and public health technologies. The Brokerage Event focused on the Region's R&D potential in biomedicine and innovative technologies for the environmental sector and public health. It featured over 100 experts, making it an ideal forum to share experience of innovative technologies and the latest developments in biomedicine. The Brokerage Event offered excellent opportunities to launch new international projects focusing on biomedicine and related technologies — which ultimately help our Region to become a major player in this exciting new field, strengthening its overall competitiveness. Experts from Ostrava's University Hospital, local universities, research institutes and companies from the Region and elsewhere gave talks on the development and practical application of biotechnologies. There were also invited speakers from the USA, Sweden and Slovakia (F. Fischer, J. Weaver and R. Stokes). As an example we can highlight the presentations about the use of new materials to treat chronic wounds, and plastic surgeons from Ostrava's University Hospital (which brings together top specialists from both the Czech Republic and abroad) spoke about their use of digital 3D technologies in reconstructing facial tissue. The main preconditions and results of the Brokerage Event were:

- Strategic Partners — The event was co-organized by RDAO and a range of partners: Moravian-Silesian Region, University. Hospital Ostrava, VŠB — Technical University Ostrava, University of Ostrava, Silesian University, and Ostrava Institute of Public Health.
- Funding — The event was funded by the Regional Authority and supported by sponsors: Molnlycke Healthcare, GE Healthcare, Biovendor and ING Corporation.
- Results — Among the new joint projects initiated as a result of the BE, there was a collaboration on stem cell research between the University Hospital Ostrava and the American company Nanoink Inc. This collaboration resulted in a US patent application.

16.2.1.2. The Brokerage Event 2012

On 19 September 2012 Ostrava hosted the Regional Development Agency's second Brokerage Event. That year's event focused on electro mobility and other eco-transport alternatives in urban areas. The Brokerage Event presented innovative projects developing electric drive systems, mainly for urban public transport applications. Natural gas-powered buses are now widespread in Czech towns and cities, but electric vehicles still tend to be associated with private cars rather than public transport. This Brokerage Event was not as successful as the previous one due to weaker topic and partner selection. Electro mobility is closely linked to the automotive industry (represented mainly by large international concerns), which is characterized by more isolated research and development activities as well as distinct supply chains. The event did not feature intensive matchmaking activities.

16.2.1.3. The Brokerage Event today

For the Moravian-Silesian Region, the Brokerage Event remains a prime route to match the results of joint projects initiated at this event and to further develop relationships between innovation stakeholders. In 2013, the Brokerage Event was devoted to the topic of "Modern materials — new possibilities for industry and healthcare". The Regional Development Agency Ostrava, as usual, was the main organizer of the event and collaborated closely with Material and Metallurgical Research Institute, Ltd. and VSB — Technical University of Ostrava, Faculty of Metallurgy and Material Engineering. The objective of the event was to share the newest findings in material research and to identify new trends in this field. The event was composed of the following three main sessions:

- Session 1: Powdered technologies of functional materials preparation.
- Session 2: Advanced metallic materials used in traumatology.
- Session 3: The use of penetration tests for evaluation of actual material properties of equipment under long term-operation.

16.2.1.4. The Brokerage Event in the future

The main benefit for the participants of the Brokerage Event is to get an overview of the organizations working in the same industry, both in the Czech Republic and abroad. There is a great possibility to find out what know-how particular organizations have and establish closer relations with them. Organizations often do not know each other and have no opportunity to establish further cooperation. This is the reason we are going to continue with organizing this kind of event each year with a main focus on the sophisticated topics from the R&D world.

16.3. Implementing best practices in your region

The Brokerage Event is a collaboration-based thematic meeting with the intent to attract the regional, national and international experts and professionals from the defined industry and create a platform for matchmaking. Each event should be based on networking between business representatives and academics and result in international collaborations in the domain of research, technology development, industrial applications or innovative entrepreneurship.

16.3.1. *Implications for organizer (e.g. TTO)*

With reference to the aforementioned, there is a need for good selection of strategic partner and for a well-set schedule. The organizing team plays a crucial role in communication with the participants of the event and in mobilization of the speakers. As stated below, the preparation process takes almost the whole year.

- Choosing a topic — recommended time for putting forward first proposals of the possible topics is ten months before the event (responsibility: organizer).
- Finding a strategic partner — recommended time for approaching the potential partner is nine months before the event (responsibility: organizer).
- Specification of the program — recommended time for tuning up the program is about six to eight months before the event (responsibility: organizer and strategic partner).
- Arranging speakers — recommended time for interesting and arranging with the speakers is five to six months before the event (responsibility: organizer and strategic partner).

- Organizing of the event — according to the deadline (responsibility: organizer and strategic partner).

Below is a sample of proposed concept of the Brokerage Event in biomedicine/biotechnology that should be outlined before arranging the event. This concept refers to the points mentioned above in the process description.

16.3.1.1. *Event organization*

The organizer will be RDAO in cooperation with their strategic partner, the University Hospital Ostrava.

16.3.1.2. *Other partners*

- VSB-Technical University Ostrava, Faculty of Electrical Engineering and Computer Science, Department of Biomedicine.
- University of Ostrava, Faculty of Medicine at University of Ostrava.

16.3.1.3. *Involvement and contribution of the University Hospital Ostrava (strategic partner)*

- Arranging for scientific foreign experts (USA, Sweden, and Slovakia) who participate in the event.
- The whole concept has to be debated with CEOs of the University Hospital Ostrava although they have confirmed to have an interest about the event.
- University Hospital Ostrava also has to preliminarily discuss the event with companies — if they are interested, willing to financially support the event etc.

16.3.1.4. *Funding*

- The cost will be approximately 20,000 euros.
- Financial Resources: Regional Authority, private partners, sponsors.

The reasons in favor of organizing the Brokerage Event on biomedicine/biotechnology:

- University Hospital Ostrava has excellent R&D results at the global level (newly awarded a US patent for the treatment of burn injuries with stem cells), is the coordinator of the project LEADER trial (phase 1 funded by FP7).
- R&D projects of University Hospital Ostrava are featured by real transfer of technologies — each R&D project has a private partner.
- The industry of biomedicine and biotechnology is getting more important in the regional economy.
- The interdisciplinary approach will be included — the event will not only be characterized by biomedicine but also by biomedical engineering represented by VSB-Technical University Ostrava, Faculty of Electrical Engineering and Computer Science, Department of Biomedicine). This approach is in compliance with the innovation policy of the European Commission which aims to promote the commercialization of R&D results across sectors of economy.
- High professional level of the event is guaranteed due to participation of leading experts from the USA, Sweden and Slovakia.
- Making visible the field of biomedicine/biotechnology as a part of the Regional Innovation Strategy and presentation of the Moravian-Silesian Region in a different light.

The most important questions that must be answered before arranging the Brokerage Event:

- Examination of the R&D results in the field of biomedicine/biotechnology in the Region.
- List of companies operating in the field of biomedicine in Region with a brief description of their business.
- Overview of companies, research institutes and universities abroad which could participate in the Brokerage Event.
- The potential cooperation among the participated organizations that should generate new R&D projects.

- Estimated number of regional and foreign participants.
- Ex-ante evaluation of the impact of Brokerage Event on the Region and its stakeholders.
- Agenda.
- Date.

As the result of the negotiations and the answers to the questions above, the event's agenda is designed. For example, the Brokerage Event agenda contained the following sessions, spread over two days:

- Presentation of the Regional Innovation Strategy of the Moravian-Silesian Region.
- European Union — current and future research subsidy programs in the field of biomedicine and biotechnologies.
- Research projects at universities and research institutes focusing on biotechnologies, environment and health — sources of fading.
- Presentation by representatives of biotechnology companies in the Moravian-Silesian Region.
- Presentation of research projects coordinated by regional and international biotechnology specialists.
- Implications for University and Research Institutions.

The universities or research institutions could provide a part of the necessary financial resources, which makes it easier for the organizer. The contribution of universities in terms of R&D results, knowledge and experience is expected. Nevertheless, the main aim for universities and research institutions should be to initiate a number of joint R&D projects from their participation to the Brokerage Event. Each Brokerage Event resulted in at least one joint project, in one case in a US patent application.

16.3.4. *Implications for policy makers*

If the organizer is not able to mobilize enough sponsors to cover all costs, it might prove necessary to utilize the public authority funds. In which case, political support is essential. The negotiations on the financial resources are considerably easier if the event is part of the innovation strategy of the region. The Brokerage Event concept is transferable to any region without any necessary modifications. However, the following requirements should be met:

- Awareness of the host organization on appropriate and necessary topics that are in the interest of potential Brokerage Event participants and that may benefit economic development in the region.
- Existence of strategic partnerships — there must be a possibility of finding a strategic partner.
- Enough financial sources — it is non-profit activity.

16.4. Future opportunities

The limitation of the Brokerage Event lies in the necessity of a good selection of the strategic partner. When an incompetent partner is selected, it is hard to attract foreign experts, which can result in a failure or in a purely regional event. In 2012 we underestimated the selection of the topic and strategic partner. Organizers should be aware that each partner has different performance motivations. Therefore, organizers should clearly analyze the partner's competence, willingness to collaborate and limitations. It is equally important to plan a schedule well in advance and insist that all partners are compliant with the terms.

- Careful selection of central topic — attractiveness of the topic is very important. This issue is closely connected with the range of experts you can engage and the openness of participants to one another.
- Better thought-out selection of participants and speakers. They must be inspired and motivated themselves and be open to further collaboration.
- Careful selection of the strategic partner — if a partner is not capable of arranging for foreign speakers, it might limit the event at the regional level.

16.5. Best practices

- Annual support and funding of the BE by the Regional Authority.
- BEs are an integral part of the Regional Innovation Strategy implementation plan.
- Established network of the strategic partners contributing to the implementation of the Regional Innovation Strategy.
- Long-standing experience with organization of professional events.

16.6. References

Papers, reports, books, conference presentations and websites

RDAO (Regional Development Agency Ostrava) (2014). Regional Innovation Strategy of the Moravian-Silesian Region 2010–2020. Retrieved from: http://www.rismsk.cz/en/download/. Accessed on: 30.06.2014.

RDAO (Regional Development Agency Ostrava) (2012a). Regional Profile. Profile of the Moravian-Silesian Region. Retrieved from: http://www.rismsk.cz/en/regional-profile/21-profile-of-the-moravian-silesian-region.html. Accessed on: 30.06.2014.

RDAO (Regional Development Agency Ostrava) (2012b). Regional Profile. Universities and colleges. Retrieved from: http://www.rismsk.cz/en/regional-profile/36-universities-and-colleges.html. Accessed on: 30.06.2014.

RDAO (Regional Development Agency Ostrava) (2012c). Regional Clusters. Retrieved from: http://www.rismsk.cz/en/regional-profile/37-regional-clusters.html. Accessed on: 30.06.2014.

RDAO (Regional Development Agency Ostrava) (s.d.). In-house documents.

Case study interviews

Interview with Brokerage Event organizing team at the Regional Development Agency Ostrava.

Interview with Deputy Director of the University Hospital Ostrava responsible for Research and Development.

17

CASE STUDY 17: The DRESDEN-concept: A Focus on Shared Services and Facilities

Nadine Schmieder-Galfe, Alexander Funkner and Oliver Uecke
(Technische Universität Dresden, Germany)

17.1. Setting the scene

Dresden has a long history as the capital of the Free State Saxony and as royal residence for the Electors and Kings of Saxony, who for centuries furnished the city with cultural and artistic splendor. Following the end of the communist-planned economy, investment was made in high-tech and the research connected to it. Therefore, Dresden exceptionally combines a long history and strong cultural scene with a world-standing research landscape. To communicate this excellent combination in a concerted way, the Technische Universität Dresden (TUD) together with non-university research institutes and cultural institutions set up a cluster titled "DRESDEN-concept". DRESDEN stands for "Dresden Research and Education Synergies for the Development of Excellence and Novelty".

Research on very basic questions benefits from interdisciplinary teams and approaches. The aim of the DRESDEN-concept cluster is to support Dresden's leading scientific areas, thereby using the synergies of all partners in terms of research, education, infrastructure and administration. The leading scientific areas include regenerative medicine, materials science, biotechnology, nanoanalysis, engineering and systems engineering. DRESDEN-concept also includes researchers from the social sciences and humanities since they study matters that are key factors of change — in politics, business, research, culture and technology (DRESDEN-concept, 2013). The idea is to go beyond basic research and translate scientific findings into innovations that are ready for the market. Dresden is a hotbed for innovative ideas originating new products or services and new start-ups. Companies like Novaled and Heliatek are just two successful examples out of many. In particular, the Fraunhofer Institutes, which focus on applied research, are a major stimulus for innovative technologies in the region. Additionally, emerging scientific fields shall be identified early on in a joint initiative to attract more world top researchers to come to the Saxon capital.

17.2. Identifying best practices

In the DRESDEN-concept cluster a central player is the TUD, which has strong partnerships with partners in research and culture areas, resulting in the ideal way to communicate the excellence of research in Dresden. In particular, the cooperation of all DRESDEN-concept partners is aimed at the development and use of synergies in research, education, infrastructure and administration (DRESDEN-concept, 2013).

The following projects are examples of how this works in practice (City of Dresden, 2013):

- Dresden Genome Center: Cross-campus technology platform providing access to the latest genome analysis methods and processes; available to all life sciences researchers on the Dresden campus.
- Technology Platform: Structure enabling equipment and infrastructure to be shared more efficiently.
- Wellcome Center:Provides visiting scientists and doctoral students from around the world with a place where they can gain information and share ideas about their stay in Dresden.

- Dresden Science Calendar: All science-related events in the Dresden area at a glance.
- Dresden Innovation Center for Energy Efficiency: Joint project for the topic of energy efficiency between the Fraunhofer Society and Dresden University of Technology.

DRESDEN-concept was founded in 2010 as an association of partners from science and culture. The 20 partners of the DRESDEN-concept network are displayed in Figure 17-1.

The association is represented by the Executive Committee, which is a set-up of rector and chancellor of TUD and the administrative director of the Max Planck Institute of Molecular Cell Biology and Genetics.

Of note, the DRESDEN-concept has been a key element of TUD's strategy as "The Synergetic University" (TU Dresden, 2013). With that strategy, TUD has successfully applied in the second round of Germany's Excellence Initiative and gained the status of being an

Figure 17-1: Members of the DRESDEN-concept consortium (DRESDEN-concept, 2013).

"Excellence University" in 2012. This Excellence Initiative was set up by the German Federal Ministry of Education and Research and aims at strengthening Germany as a science and research location, increasing the visibility of cutting edge research at German institutions of higher education, and improving Germany's international competitiveness. All partners have been members of the DRESDEN-board, which acted as advisory board for the rector during the application compilation time for the Excellence Initiative.

17.2.1. *Specific description of shared facilities ("Technology platform")*

Within the DRESDEN-concept consortium, a project called "technology platform" has been set up, which aims at mapping available technologies, devices and services of the individual partners in a free-of-charge online platform. That allows both partner employees as well as external interested parties a detailed and cross-institutional overview of the research infrastructure in Dresden. With that, even small research groups get access to expensive devices, which they themselves could not afford to buy. As a positive side effect, the platform also enhances collaborations, which are especially important in the life sciences field and an essential precondition to being a successful researcher and to creating an overall vibrant research environment.

The user of a technology pays a user fee, part of which is returned to the provider as a revenue which he/she can in turn use to finance own research activities. The platform allows the owner of an expensive device to utilize it to capacity, and therefore maximize the cost efficiency of the usually significant capital expenditure. To protect the provider from misuse of the device and the user of a certain technology from unexpected claims during or after usage, both parties agree on specific terms of use.

In terms of content, the technology platform provides detailed information on:

- scientific devices and their technical characteristics
- scientific range of services and their according modalities

- scientific technologies and methods
- expert profiles that describe scientific competences available in the facilities.

17.2.2. Evolution of the DRESDEN-concept for shared facilities

The technology platform was inspired by the Max Planck Institute of Molecular Cell Biology and Genetics (MPI-CBG) that represents the major pioneer in optimizing research infrastructure by centralizing a broad range of scientific services and facilities, which are on offer to internal and external customers. Of note, the MPI-CBG also provides methodological guidance from experts that head the different departments. Services and Facilities that are provided at the MPI-CBG are listed in Figure 17-2.

This concept of shared services and facilities has also already been implemented at the BIOTEChnological Center of TUD and the Center for Regenerative Therapies Dresden (CRTD) — another major driver of the success of TUD's application within the Excellence Initiative.

17.2.3. The "Dresden Model: Shared facilities" today

So far, the following nine partners of the DRESDEN-concept association provide their equipment, including detailed information on classification and technology, on the online platform http://tp.dresden-concept.de: the TUD, the University Clinics Carl Gustav Carus Dresden, the MPI-CBG, the Fraunhofer-Institut für Keramische Technologien und Systeme (IKTS), the Fraunhofer Institute for Material and Beam Technology, the German Center for Neurodegenerative Diseases, the Helmholtz-Zentrum Dresden-Rossendorf, the Leibniz Institute for Solid State and Materials Research and the Leibniz Institute for Polymer Research Dresden. The number of available technologies can be seen in Figure 17-3, which depicts a screenshot of the online platform.

Services & Facilities at the MPI-CBG

The following table contains all the services and facilities at the MPI-CBG as well as their service leaders. To get to a services and facilities' page simply click on the name or the service leader of the facility. You can also use the search form at the bottom of the table, to search for services and facilities or service leaders within the table.

Services & Facilities	Service Leaders
Antibody Facility	Patrick Keller
Bioinformatics	Ian Henry
Biomedical Services	Jussi Helppi
Budget Office	Ralf Meier
Chromatography	Barbara Borgonovo
Computer Services	Matt Boes
DNA Sequencing	Sylke Winkler
Electron Microscopy	Jean-Marc Verbavatz
FACS	Ina Nüsslein
High-Throughput Technology Development Studio (TDS)	Marc Bickle
Information Office	Florian Frisch
International Office	Carolyn Fritzsche
Laboratory Management	Deborah Newby
Library	Silke Thüm
Light Microscopy	Jan Peychl
Mass Spectrometry	Anna Shevchenko
Microarray Analysis	Julia Jarrells
Photolab	Franziska Friedrich, Kostas Margitudis
Protein Expression and Purification	David Drechsel
Office of Research Grants	Birgit Knepper-Nicolai
Office of Technology Transfer	Ivan Baines
Tilling	Sylke Winkler
TransgeneOmics	Mihail Sarov
Transgenic Core	Ronald Naumann

Figure 17-2: Services and facilities offered at the MPI-CBG (MPI-CBG, 2013).

The most recent version of the online platform offers data maintenance and search functions for scientific equipment, either directly via the website or through using selected search engines. To implement their own equipment, potential providers need to get authorization. This is currently set up in a decentralized manner, i.e. each institution rather than the DRESDEN-concept people will provide the platform user with the appropriate data. Pre-existing databases like the Research Information System of TUD (which contains among other things information on available research infrastructure), or the research infrastructure databases of the MPI-CBG, CRTD and BIOTEC, are connected directly with the technology platform. Thus,

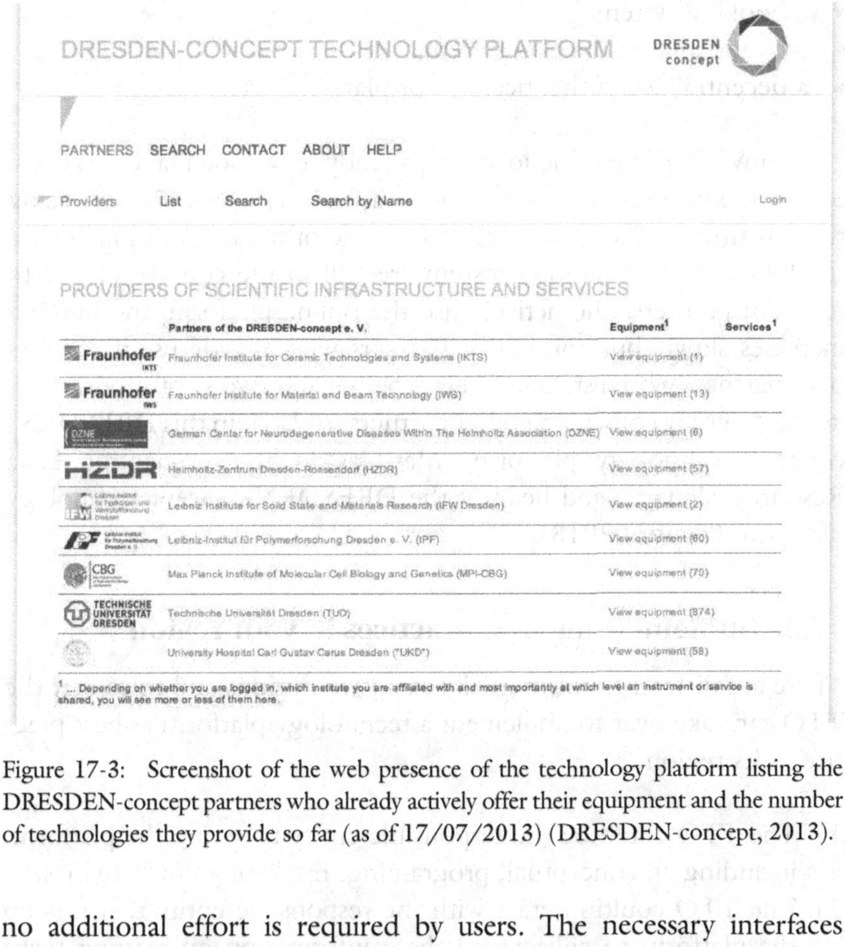

Figure 17-3: Screenshot of the web presence of the technology platform listing the DRESDEN-concept partners who already actively offer their equipment and the number of technologies they provide so far (as of 17/07/2013) (DRESDEN-concept, 2013).

no additional effort is required by users. The necessary interfaces between the different databases and the online platform are already developed (Matthias Fichtner, Scientific Data Resource Manager and head of the DRESDEN-concept technology platform, 09 April 2013).

17.2.4. The DRESDEN-concept with shared facilities in future

Summarized, functions that will be implemented within the technology platform include (Fichtner *et al.*, 2011; Fichtner, 2012):

- data maintenance and data synchronisation
- various data search tools

- a booking system
- a billing and accounting system
- a decentralized authentication for platform users.

Many of these functions are already developed and available online. Some users can already gain access via single sign-on using their institutional login for example. Some of the functions including the billing and accounting system are still in the concept phase. In terms of partners who actively use the online platform, the number increases slowly but constantly. Partners who already use it serve as pilot partners who distribute their experience by word-of-mouth marketing, thereby convincing other partners to also join this DRESDEN-concept technology platform (Matthias Fichtner, Scientific Data Resource Manager and head of the DRESDEN-concept technology platform, 09 April 2013).

17.3. Implementing best practices in your region

There are different, not mutually exclusive options and roles that the TTO can take over to implement a technology platform as best practice in its region:

1) The TTO could be the responsible entity to set up the platform, including all conceptual, programing, marketing etc. activities.
2) The TTO could interact with the responsible entity that sets up the platform to collect and share information on existing technologies and devices, to advertise and to convince researchers to join the platform.
3) The TTO could use the online platform as an information source to provide knowledge about available technologies and devices or potential collaborations between researchers and between researchers and the industry. In particular, the TTO could use this source to inform spin-offs of the institution to foster their activities and growth, and to strengthen the connection of the spin-off with its "mother" institution.

A top-down directive towards the researcher to provide especially expensive or rarely used equipment and devices to others is the utmost important task of the university or research organization to successfully implement a technology platform as broadly and therefore as usefully as possible within a region. Ideally, many research institutions combine their forces to maximize synergies. This network should be set up through the lead of the university or research organization, which would need to get a real commitment from all institutions to convince their researchers to use this platform. Researchers of this "lead institution" could become pilot users of the platform, which would in turn trigger further users of other institutions.

Policy makers need to provide money to pay a team who sets up a technology platform. Besides programing, collecting information, marketing activities to gain users etc., that team would also take over the task to connect many local institutions and their facilities and technologies. Furthermore, policy makers could adjust the legal framework to simplify charging of utilized technologies, for example with regards to tax issues like VAT. A rather radical, yet increasingly necessary top-down task is required from policy makers of national funding institutions or ministries towards researchers and scientific institutions: recipients of public equipment grants should be forced to share expensive equipment/devices (e.g. more than one million euros) to reach a 100% utilization rate. This would inevitably lead to the setup of technology platforms and therefore to an increase in efficiency in achieving new scientific results for a particular amount of granted public research funds.

Of note, the concept of technology platforms is starting to evolve in research institutes all around the world. In a future scenario, it is not unlikely that scientists will only work in a research environment of shared technologies and services to cope with the increasing complexity of science, research approaches and the constantly increasing number of available technologies and devices.

17.4. Future opportunities

- Increase number of partners to ideally provide a complete list of scientific technologies and devices within the Dresden research community.
- In future, centralize part of the researcher's revenues as overheads to pay technology experts who provide guidance on experimental design, efficient use of technologies etc. towards the researchers.

17.5. Best practices

- Comprehensive overview of a broad range of technologies available in one area facilitates scientific exchange within and between scientific institutions.
- Easy and free access to information via online.
- Technology experts provide guidance on experimental design, efficient use of technologies etc. towards the researchers, and therefore prevent long self-educating replication cycles of experiments.
- Allows maximizing the use of currently not fully occupied devices that increases cost-efficiency and allows in future additional revenues for the owners of those devices.

17.6. References

Papers, reports, books, conference presentations and websites

City of Dresden (2013). DRESDEN-concept. Retrieved from: http://invest.dresden.de/de/Wissenschaft_Innovation/DRESDEN/concept_1267.html. Accessed on 15.07.2014.

DRESDEN-concept (2013). Retrieved from: http://www.dresden-concept.de. Accessed on 15.07.2014.

Fichtner, M., Mahn, T., Odenbach, M. & Köhler, T. (2011). Kosteneffizient forschen. Dresdner Transferbrief. 3.11, p. 17.

Fichtner, M. (2012). Dresdner Technologieplatt form wächst. Newsletter HZDR "Insider". 06/2012, p. 5.

MPI-CBG (2013). Facilities. Retrieved from: http://www.mpi-cbg.de/facilities. Accessed on 10.10.2013.

TU Dresden (2013). Die Synergetische Universität. Zukunftskonzept zum projektbezogenen Ausbau der universitären Spitzenforschung.

Case study interview:

- 09.04.2013, Dr. Matthias Fichtner, Scientific Data Resource Manager and head of the DRESDEN-concept technology platform, Dresden.

Conclusion

This book presents 17 best practice case studies on effective technology transfer in biotechnology. In this main conclusion, we present an overview of the major results and outcomes of the different sections. Following this, we integrate the main conclusions of the whole book.

Technology transfer offices

TTOs are crucial intermediates between the academic and industrial world. Case studies on two well-established TTOs reveal that there are different ways in which these offices position themselves in relation to external (government, industry, etc.) and internal stakeholders (scientists, institute management, etc.). The examination of a newly formed TTO highlights how initial vision and strategy of a TTO define its role and future success path. We observe that a young TTO can — and probably should — act relatively pragmaticly while strategies, procedures and infrastructure is already well established in more experienced TTO structures. We also saw that there are as many TTO models as there are TTOs. Some TTO structures employ innovative business models. This is for instance the case in Germany where universities cannot be the shareholder of their spin-offs.

TTOs try to increase their efficacy and efficiency by specializing in specific technology fields and by putting in place professional processes and structures. The selected case studies describe two traditional non-profit TTOs (VIB and IICMB) and two for-profit TTOs (TU Dresden and Imperial). However, we have shown that both models can be effective. Therefore, we conclude that TTOs are heavily influenced by their external and internal environment, especially by policy makers and of course by the scientific community they are servicing. It appears to be that government commitment and a long-term regional vision on R&D are necessary preconditions for TTOs to thrive and to develop.

Funding

This section focuses on how technology transfer is funded in different parts of Europe. There appears to be as many funding mechanisms to support spin-off and licensing activities as there are cases in this book. One way of approaching the funding gap is that universities, and more specifically TTOs, act as sort of private investors that use less strict milestones to support spin-offs to bridge the funding gap in the early stages. The TTO then manages its own investment fund and in some cases also participates in investment rounds beyond the early stages of a spin-off.

The government can offer financial tools to TTOs and researchers to facilitate and financially support technology transfer. We have seen that this can be done by either structurally supporting applied research in universities and research institutes or by stimulating industrial partners to actively seek collaboration with the regional scientific community. We have also presented the case where this support to the technology transfer process is centrally organized. Different programs are explained that provides financial resources for patent protection, public R&D projects, R&D centers and special purpose vehicles that act as intermediaries between research institutes and the market. All case studies, again, underline the importance of a strong government commitment towards innovation and entrepreneurship.

Incubators

Universities have the crucial task to protect, support and facilitate the early stages of their spin-offs. One way of doing this is by making sure that there are suitable incubators in place in the region. Incubators are structures that provide start-ups with office space, lab space and business services. They have proved to be highly successful structures in supporting biotechnology companies and in promoting technology transfer between universities and industry. In this section, we discuss the best practices in founding and developing incubators, science parks and accelerators. Of crucial importance is the incubator's design and its role in facilitating network activities and an entrepreneurial community. Incubators should collaborate closely with surrounding TTOs and have been shown to be able to create an idea development program for bachelor and master's students. These students can form an important source of new ideas, concepts and prototypes. We should indeed point out that a lot of interesting ideas and concepts are developed by non-academic staff and are not high-tech and sometimes not protectable, something that can be cured by involving students in the incubator operations.

Entrepreneurship education

Scientific staff members are at the basis of new discoveries, technology platforms and knowledge. They are often the driving force behind technology transfer projects, so training and educating these people in the fields of entrepreneurship and technology transfer is regarded to be one of the key solutions to fostering technology transfer. Central to most educational approaches is "cross-fertilization" between business and research institutes and between students from different faculties or scientific disciplines. But developing and implementing this training or curricula is more than setting up a number of basic courses. Equally important are the setting up of activities that create and maintain an entrepreneurial culture within the organization and providing coaching to scientists or entrepreneurial teams. Entrepreneurial expertise, something that is often not readily available at research institutions, should be brought to the table.

In this book, we have presented best practices of comprehensive educational programs at academic organizations that stimulate entrepreneurship and technology transfer. This section also introduces a successful training and coaching program in the region of Catalonia, Spain that supports and trains entrepreneurial teams during the creation of their life sciences company. This best practice emphasizes the importance of a "learning by doing" approach to technology transfer education.

Clusters

Technology transfer projects are aided or hindered by their environment. In this book, we have looked at a number of so-called biotechnology clusters and how these high concentrations of interconnected biotechnology organizations can impact regional technology transfer. Biotechnology clusters start and grow in the vicinity of high-quality research lab, attracting more activities, investors and interest to the region. At a certain point, the networks, culture and knowledge flows within a cluster are such that they create a self-sustaining momentum: the cluster reaches a critical mass and can grow further. In this book, we have seen how a number of regions have established or try to establish a biotechnology cluster. Invariably, the focus of these initiatives is on bringing high-quality science/scientists into contact with the external environment. This implies open and intensive collaboration between the academic, industrial and governmental stakeholders. We have seen that this type of collaboration and the informal networks emerging in a biotechnology cluster can have a beneficial influence on the effectiveness of biotechnology spin-off projects. We show that supporting clustering activities is more than supporting formalized collaboration or scientific research: it is about building a community and about long-term support of an environment in which valuable ideas and initiatives can germinate and grow.

General conclusion

First and above all, high-quality research lies at the heart of technology transfer. If there is no competitive, differentiating research activity, there are no basic inputs for technology transfer. Once we go past this basic assumption, we observe that initiatives and measures are taken in a wide variety of domains. The best practice case studies selected for this book focused on the core organizations involved in technology transfer in biotechnology: technology transfer offices, business incubators and research institutes. We then looked at funding mechanisms on different levels that facilitate the transfer of know-how and technologies from academia to industry. In a last step, we consider how technology transfer activities are embedded within a regional ecosystem.

This is a very broad scope for a handbook. However, it is clear that the effectiveness of technology transfer can only be improved by looking at technology transfer as a multi-faceted phenomenon. Technology transfer results will not be boosted or grow rapidly by taking a couple initiatives or by focusing investments or efforts on one domain. In this book, we have shown that improving technology transfer effectiveness requires action and long-term commitment on all levels. Governments should be willing to invest a lot of resources and effort for a long period. Institutions should change their organizational structure and processes; and possibly rethink their reward systems. Scientific laboratories should collaborate and negotiate with industrial partners; activities for which a new set of skills and capabilities should be attracted to the lab. Individuals should be made aware that getting involved in technology transfer is a long term, sometimes frustrating, but always rewarding opportunity.

In recent years, a lot more attention is put on how and if academics provide societal impact as a result of their receiving R&D funding. Technology transfer provides one route of returning discoveries and technologies to society, but it is one that requires a lot of upfront investment, intensive collaboration, vision, an all-encompassing strategy and long-term commitment.